83 Notes on Numerical Fluid Mechanics and Multidisciplinary Design (NNFM)

Springer

Berlin
Heidelberg
New York
Hong Kong
London
Milan
Paris
Tokyo

Engineering ONLINE LIBRARY

http://www.springer.de/engine/

LESFOIL: Large Eddy Simulation of Flow Around a High Lift Airfoil

Results of the Project LESFOIL
Supported by the European Union 1998 – 2001

Lars Davidson, Davor Cokljat, Jochen Fröhlich,
Michael A. Leschziner, Chris Mellen and Wolfgang Rodi
(Editors)

Springer

Prof. Dr. Lars Davidson
(coordinator)
Dept. of Thermo and Fluid Dynamics
Chalmers University of Technology
SE-412 96 Gothenburg, Sweden

Dr. Davor Cokljat
Fluent Europe Ltd.
Sheffield Airport Business Park
Europa Link
Sheffield S9 1XU, Great Britain

Dr. Jochen Fröhlich
Dr. Chris Mellen
Prof. Dr. Wolfgang Rodi
Institute for Hydromechanics
Universität Karlsruhe
Kaiserstr. 12
D-76128 Karlsruhe, Germany

Prof. Dr. Michael A. Leschziner
Imperial College of Science
Technology and Medicine
Aeronautics Department
Prince Consort Rd.
London SW7 2BY, Great Britain

ISBN 3-540-00533-1 Springer-Verlag Berlin Heidelberg New York

Library of Congress Cataloging-in-Publication-Data
LESFOIL : large eddy simulation of flow around a high lift airfoil : results of the Project LESFOIL supported by the European Union 1998 - 2001 / Lars Davidson ... [et al.] editors. p. cm. --
(Notes on numerical fluid mechanics and multidisciplinary design ; 83)
Includes bibliographical references.
ISBN 3-540-00533-1
1. Aerofoils. 2. Lift (Aerodynamics)--Research--Europe. 3. Project LESFOIL
European Commission) I. Davidson, Lars. II. Series.
TL574.A4L47 2003
629.134'32--dc21

Springer-Verlag Berlin Heidelberg New York
a member of BertelsmannSpringer Science+Business Media GmbH
http://www.springer.de

Printed in Germany

Typesetting: camera-ready by authors
Cover design: deblik Berlin
Printed on acid free paper 62/3020/M - 5 4 3 2 1 0

NNFM Editor Addresses

FOREWORD

Large Eddy Simulation is a relatively new and still evolving computational strategy for predicting turbulent flows. It is now widely used in research to elucidate fundamental interactions in physics of turbulence, to predict phenomena which are closely linked to the unsteady features of turbulence and to create data bases against which statistical closure models can be assessed. However, its applicability to complex industrial flows, to which statistical models are applied routinely, has not been established with any degree of confidence. There is, in particular, a question mark against the prospect of LES becoming an economically tenable alternative to Reynolds-averaged Navier-Stokes methods at practically high Reynolds numbers and in complex geometries. Aerospace flows pose particularly challenging problems to LES, because of the high Reynolds numbers involved, the need to resolve accurately small-scale features in the thin and often transitional boundary layers developing on aerodynamic surfaces. When the flow also contains a separated region – due to high incidence, say – the range and disparity of the influential scales to be resolved is enormous, and this substantially aggravates the problems of resolution and cost. It is just this combination of circumstances that has been at the heart of the project LESFOIL to which this book is devoted.

The project combined the efforts, resources and expertise of 9 partner organisations, 4 universities, 3 industrial companies and 2 research institutes. These partners collaborated closely over a period of 3 years, addressing a wide range of issues, both numerical and physical, in an effort to answer the question posed above in relation to the utility of LES for airfoil flows. Hundreds of simulations were undertaken, many hours of sometimes heated discussion took place and many weeks have been devoted to poring over results, understanding their implications and removing inconsistencies through repeated simulations and critical inquiry. For the large majority of researchers involved, LESFOIL has been a rewarding and fruitful experience, well-worth the effort and expense. The LESFOIL activities were co-funded by the Fourth Research Framework Programme of the European Union (EU).

All partners deserve recognition and gratitude for their commitment, hard work and willingness to share their knowledge and experience. Special appreciation must be given to the European Commission, notably to Dr. Dietrich Knörzer, for recognising that a still speculative, somewhat fundamental strategy deserves to be promoted and funded alongside the usual industrially-oriented projects which are the backbone of the EU's R & D programme. Finally, the editors would like to thank Prof. E.H. Hirschel, General Editor of the Springer Series Notes on Numerical Fluid Mechanics', who made it possible to publish the outcome of LESFOIL in the present book.

Maj 2002
LD, DC, JF, MAL, CPM, WR

Table of Contents

I. INTRODUCTION

I. INTRODUCTION

LES is widely assumed to be destined to replace conventional RANS modelling as the preferred predictive approach to certain types of engineering flows. This expectation is rooted in the observation that simulations yield decisively superior solutions, even for gross time-mean properties, for flows that feature periodic motion and/or in which the turbulence structure is dominated by large-scale eddies, typically of the order 0.1-1 of the appropriate dimension of the flow itself. This is the case, for example, in bluff-body flows and complex free-shear layers in which the mean features are, essentially, insensitive to the influence of walls, at least in terms of viscosity-affected processes. In such circumstances, turbulence transport is dictated by the unsteady dynamics of the interactions between the flow and the large-scale structures and amongst the various periodic and turbulent scales. The statistical consequences of this state of strong departure from structural equilibrium are extremely difficult to represent realistically with RANS models.

Although the cost of LES is much higher than that of RANS – typically by a factor 50-500, depending on geometry and type of flow – LES is gaining ground as a result of rapid computer-hardware developments, yielding a doubling of computing power roughly every two years. Moreover, unlike DNS, the resolution requirements of LES for flows that are not sensitive to near-wall processes rise only weakly with Reynolds number (typically with $Re^{1/2}$), and the contribution of subgrid-scale transport to the corresponding resolved statistical properties is subordinate. On the other hand, the accuracy of LES is quite sensitive to grid quality, ie. aspect ratio, skewness and rate of change of internodal disctance. Thus, in complex geometries, the accuracy of LES can be poor even in regions that are remote from walls.

LES is confronted with much greater challenges in near-wall flows. As the wall is approached, the large scales progressively diminish in size, eventually approaching the Kolmogorov scale. In addition, the near-wall structure is highly anisotropic, characterised (in shear flow) by elongated vortical structures of dimensions $(x^+, y^+, z^+) = \mathcal{O}(100, 10, 20)$. Hence, the grid supporting the near-wall layer must approach one that would normally be used in a DNS. The requirement that the grid-aspect ratio be constrained to accommodate the above structural features of the near-wall motions quickly leads to economically untenable grids as the turbulence Reynolds number increases beyond 1000.

Flows which separate from curved surfaces are especially difficult to predict, by whatever computational strategy. In the most taxing conditions, separation is approached gently, following a progressive deceleration and, in the case of three-dimensional flows, skewing of the converging boundary layers. In such circumstances, the structure of the near-wall layer, in particular its response to adverse pressure gradient and wall-related constraints, is highly influential and must be represented accurately. This is the case in high-lift airfoils that approach maximum lift and then breach this boundary to provoke stall. Such conditions are of principal interest to wing designers who are concerned with predicting lift and drag in take-off and especially landing conditions. With separation control becoming an increasingly important practical topic – one major aim being the elimination of the leading-edge slat and (some) flaps – high accuracy in the prediction of separation is a key requirement in aerospace.

The performance of RANS models in high lift conditions has been the subject of much attention over the past decade. This is exemplified by compilations produced by Haase *et al* [1, 2] and Dervieux *et al* [3] all arising from European validation initiatives. Configurations receiving particular attention were the single-element Aerospatiale-A and NACA 4412 airofoils and the two-element NLR 7301 airfoil. Fundamentally associated with these cases is the much more complex case of a prolate spheroid at high incidence [4–6], with massive separation occurring on its leeward side. A wide variety of turbulence model were tested for these cases, and the studies demonstrated the serious difficulty of identifying a single closure framework that could be trusted to give adequate predictive accuracy for all important conditions, especially at incipient stall. It is against the background of this experience, and progress of LES in non-aerospace sectors that has motivated the LESFOIL project to which this book relates.

The overall aim of LESFOIL was to identify the potential of LES as a prediction method for separation in high-Reynolds-number airfoil flows, representative, in a generic sense, of unswept, high-aspect-ratio wing elements at high incidence. The emphasis of the project was, intentionally, on the fundamental issues encountered in (operationally) "simple", nominally two-dimensional stall, rather than on complex practical geometries, for which there are, in any event, hardly any experimental field data.

The above aim was addressed through an international partnership of the nine academic and industrial groups listed below:

Chalmers University of Technology, Sweden (coordinator);
Alenia Aerospazio, Italy;
CERFACS, France;
Dassault Aviation, France;
Fluent Europe, UK;
University of Karlsruhe, Germany;
ONERA, France;
University of Surrey, UK;
UMIST (Manchester), Queen Mary College (University of London), Imperial College (University of London), UK.

The involvement of three universities under the last item in the above list is a consequence of the associated research group moving between universities during the course of the project. The predominance of academic and research-institute groups in the partnership reflects the largely generic nature of the research undertaken in LESFOIL, with emphasis being placed on geometrically simple flows, allowing fundamental issues to be investigated.

In view of wide-ranging experience in earlier collaborative EU studies on RANS approaches, the particular airfoil geometry selected was the Aerospatiale-A airfoil [7], operating at a chord Reynolds number of $2.1 \cdot 10^6$, angle of incidence 13.3^o and Mach number of 0.15. The airfoil geometry is shown in Fig. 39 at p. 75, while the detailed flow conditions around it will be reviewed in Chapter III (the experimental data as well as the geometry of the airfoil can be found on CD-ROM in [2]). Key factors leading to this choice were, first, the nominally two-dimensional nature of the flow; second, the availability of detailed data obtained in two wind tunnels; and third, the availability of many RANS solutions, allowing comparisons to be made with the LES results. The first feature, in particular, permitted a restriction of the spanwise extent of the computational domain and offered the opportunity to answer important questions on the influence of varying the spanwise extent and the resolution therein on the realism of the simulated flow.

At the heart of LESFOIL was the question as to what practices are appropriate (or, indeed, inappropriate) for the resolution of the boundary layer on the suction side of the airfoil. Apart from the occurrence of separation and the extreme thinness of the semi-viscous sublayer at the high Reynolds number, additional complicating features of the A-Airfoil flow included the presence of leading-edge separation and free transition at around 12% of chord on the suction side. Thus, despite the geometric simplicity of the airfoil geometry, the flow around it is extremely challenging and poses a number of important questions about the applicability of several elements contributing to LES. In particular, LESFOIL set out to investigate the effectiveness of using a range of approximate near-wall treatments based on wall-functions and RANS-LES hybrids wherein a RANS method is applied in the near-wall region and coupled to the outer LES domain.

While much of the total effort within LESFOIL was directed towards the airfoil flow, a whole range of precursor studies were undertaken to investigate key fundamental facets pertinent to decisions on the adoption of a limited range of practices for subsequent airfoil simulations. These studies have been grouped under the following four headings:

Characteristics of subgrid-scale models
Near-wall modelling for attached and separated flows
Numerical practices for LES
Airfoil simulations

The above sequence is broadly in line with that of major "tasks" or research strands pursued within LESFOIL. Thus, in an initial study, a number of computational approaches were examined, with the objectives being of identifying relative accuracy and computational efficiency and any "best practices" which might be shared and adopted by collaborating contributors. The next phase focused on subgrid-scale modelling. A wide range of model variants were investigated, mainly by reference to channel flows, to identify the most promising approaches for subsequent separated-flow simulations. In this context, particular attention was directed towards the near-wall characteristics of alternative models, especially their asymptotic behaviour.

A major study preceding that of the airfoil simulations was one in which four LES schemes were applied to separation from a curved hill-shaped surface within a streamwise periodic channel segment at a Reynolds number, based on channel height, of 22,000. The geometry is shown in Fig. 9 at p. 22. At this relatively modest Reynolds number, almost fully-resolved simulations could be performed with a fine grid of close to 5 million nodes, and these provided accurate benchmark data against which corresponding simulations, undertaken on coarser grids with a range of approximate near-wall and subgrid-scale models, could be compared. The principal objective of this study was to derive conclusions on which near-wall practices should be adopted for subsequent airfoil simulations for which wall-resolved grids were assumed to be untenable on cost grounds. The details of this study and its outcome are documented in Chapter II.

The final and principal part of LESFOIL was that concerned with airfoil simulations. In total, some 70 simulations were undertaken for this case with a range of grids, numerical practices, near-wall approximations and subgrid-scale models. Chapter III and IV presents a subset of "best" results and discusses the major findings emerging from the extensive investigation. A crucially important aspect of the study was a close interaction and collaboration between contributors in terms of common grids, numerical practices and modelling issues. This entailed an "iterative" approach in which simulations were repeated, in some cases several times, and also extended in terms of the range of parameters studied, in an effort to eliminate inconsistencies and arrive at a consensus view on the capabilities of LES for separated airfoil flow. This view is conveyed and justified in Chapter IV.

II. PREPARATORY WORK

II. PREPARATORY WORK

1 Task 1: Subgrid models

Task Leader: CERFACS
F. Ducros

1.1 Summary of work progress

This task consists in preliminary works, that prepare the airfoil simulation. It covers the mesh generation (Task 1.1), as well as some investigations of LES capabilities with respect to both physical and numerical parameters of importance for the simulation of the flow around the Airfoil. The ability of subgrid scale models in representing transition, near-wall region, flow separation is studied. Investigated numerical parameters include grid resolution, grid distortion and of course numerical schemes. These issues are covered by Task 1.3. For the particular case of the channel flow and to provide accurate references, a DNS data base has been produced (Task 1.2).

1.2 Task 1.1: Grid generation

The subsonic A profile of Aérospatiale at incidence 13.3^o, Mach number 0.15 and Reynolds number $2.1 \cdot 10^6$, has been chosen as the final LES test-case of the project. Documented experimental data as well as numerous steady numerical investigations are available. Moreover, the low Mach number allows realistic use of incompressible flow solvers, while compressible flow solvers can still be used.

Four meshes have been provided, including two topology types (structured and unstructured) and two level of wall-resolution: a *gold* mesh for standard no-slip conditions, $y^+ = 1$ in the wall-normal direction, and a *Wall function* grid, dedicated to the use of wall function, with $y^+ \approx 100$. Even if some simulations could be performed using other meshes, these shared meshes are planned to constitute a reference for all partners.

Unstructured grids. The *Gold* (resp. *Wall function*) grid contains 441888 (269088) points, 2552013 (1547241) tetrahedral elements and 7750 (7750) wing surface triangles. This mesh was generated by Dassault.

Structured C-grids. The 3D mesh consists in equidistant planes of the same 2D mesh. With 32 cells in the span, the *Gold* grid contains $512 \times 96 \times 32 = 1572864$ cells and the *Wall function* grid $256 \times 64 \times 32 = 524288$ cells. Both grids are well suited for multigrid algorithms, with power of 2 number of cells in the various regions (see Fig. 1). This mesh was generated by ONERA.

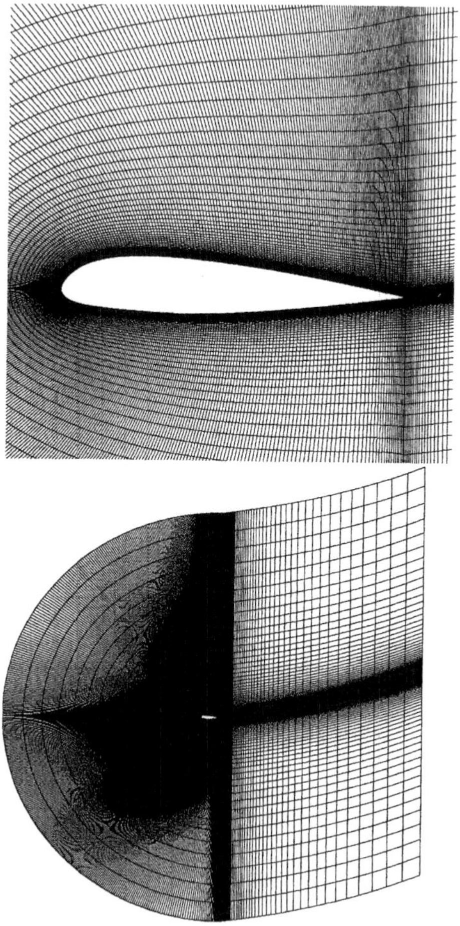

Figure 1 C-structured grid around the A-airfoil profile. Close view (top) and global view (bottom).

1.3 Task 1.2: Generation of database with DNS

Prior to performing LES, a Direct Numerical Simulation (DNS) of the turbulent channel flow at $Re_\tau = 180$ was performed by CERFACS, constituting a database for a-priori testing of subgrid-scale models. This DNS was validated by detailed comparisons with other published results [8], see Fig 2.

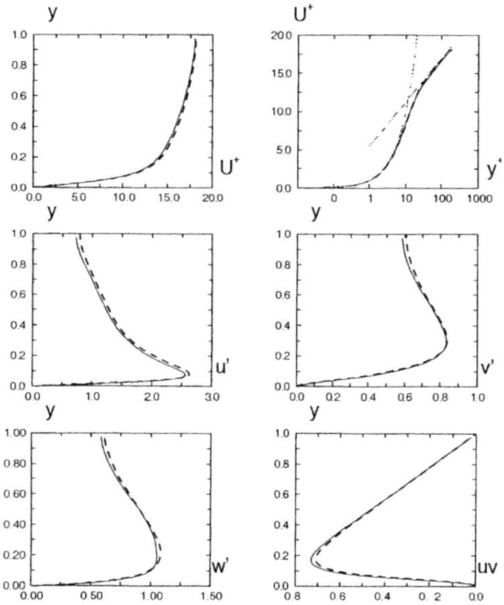

Figure 2 Cerfacs: DNS at $Re_\tau = 180$. Profiles of mean velocity, rms velocity fluctuations (u', v', w') and Reynolds shear stress. Solid lines: present DNS; dashed line: DNS from [8].

1.4 Task 1.3: Development and evaluation of subgrid models in simple configurations

This sub-task is the core of Task 1. It aims, through several test cases (see Table 1) performed with varying parameters to put modelling procedures in situations where they encounter features typical of airfoil flow simulations (see table 2). The key-words of the investigations are model robustness and applicability to general complex flows.

The channel flow is the basic test-case for wall-bounded flows. Due to its simplicity it allows extensive testings. The hill geometry [9, 10] has been selected as a test case of flow representative of separation on smooth profile and backflow (see Fig. 3, top). The flow around a surface-mounted cube is a standard test-case [11, 12] for both RANS and LES methods (see Fig. 3, bottom). For both cases, the flow topologies are complex, involving multiple separation regions and vortices.

Complementary work on crucial features of subgrid-scale models, *i.e.* their behaviour in the near-wall region and their ability to handle with transition, is described in more details in Task 2 (Near-Wall modelling) in Section 2.

Table 1 Academic flow configurations investigated during Task 1.3: × means that the corresponding partners has treated the configuration.

Flow Configuration	Chalmers	CERFACS	Fluent-E	Karlsruhe	Surrey	UMIST
Channel	×	×	-	×	×	×
Hill	×	-	-	×	-	×
Cube	-	-	×	-	-	-

Evaluated subgrid-scale models. Only eddy-viscosity models have been considered. The classical Smagorinsky model with various wall damping functions is for every partner the reference model. However, this model has a well-known deficiency: it is sensitive not only to resolved turbulence but to every velocity gradients, including mean flow or large scale gradients. This makes it inaccurate, especially in wall bounded turbulent flows and also unable to predict transition to turbulence. The standard structure function model [13] has similar deficiency. More elaborate models, able to detect small-scale turbulence in a flow field, have thus been tested.

Several partners have used the dynamic Smagorinsky model. This model proved to be accurate at low and moderate Reynolds numbers but it requires an averaging procedure to remain stable. This prevents its use in complex flow. The localized dynamic model [14] overcomes this difficulty, as well as the one-equation dynamic model proposed by Chalmers [15]. In this latter model, a transport equation for the subgrid-scale kinetic energy is introduced which provides in a smooth way the model with all dynamic informations. Dynamic models can mimic backscatter energy transfer.

Other tested models are the filtered structure function model with and without wall adaptation [16], the WALE model proposed by CERFACS [16] and the Mixed Scale Model proposed by ONERA [17]. These models are

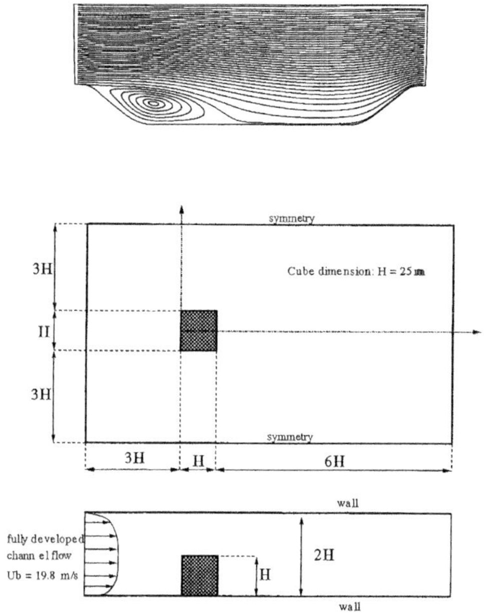

Figure 3 Topology of the flow for the hill flow (top) and global configuration of the "flow around a cube" test case (bottom).

completely localized. They also act selectively on small scale turbulence but contrary to dynamic models, they include constants that should be adjusted.

Table 2 Flow, numerical and modelling parameters investigated in Task 1.3 ("stretching" relates to investigation of grid distortions effects, WD stands for Wall Damping, WF for Wall function, and "regime" means that various Reynolds numbers are considered or that transition is studied - see text for details).

Flow Configuration	models	resolution	stretching	WD	numerics	regime	WF
Channel	×	×	×	×	×	×	×
Cube	×	×	×	-	×	-	×
Hill	×	×	-	-	-	-	×

Most of the models treat the direction of mean flow gradients (wall normal direction in wall bounded flows) and other directions separately. This may be not too much a problem for the airfoil simulation since the wall normal direction is clearly identified but limits their use for general application. For example, wall damping applied to the Smagorinsky model or wall adaptation of structure function models require knowledge of the wall normal direction; also dynamic procedures usually rely on 2D test filters applied in planes parallel to the wall. Models explicitly including the grid size as the length scale in their definition (all models except dynamic models for which the length scale is evaluated dynamically together with the constant) may also loose accuracy in high gradients regions where grid is usually strongly anisotropic.

On the whole, eight eddy-viscosity models have been tested. These investigations have allowed to check numerous already-known results but also to initiate new studies and draw some original conclusions. Below the models are discussed in relation to resolution, wall damping, numerics, Reynolds number regime and wall functions.

Resolution. For the channel flow test-case, a large range of resolutions has been used, from well resolved LES with $\Delta y^+ \approx 1$ at the wall and $\Delta x^+ \times \Delta z^+ \approx 30 \times 10$ in the streamwise and spanwise directions to more aggressive grids ($\Delta y^+ = 2$ at the wall and $\Delta x^+ \times \Delta z^+ \approx 400 \times 150$), more representative of the mesh (*Gold mesh*) around the airfoil.

In any case, the simulations are able to provide realistic results, even for very crude resolutions, showing that the selected models are robust. For most of the tests, the results improve as the mesh is refined, which is intuitive. Along with this conclusion, it is shown that the SGS models cannot palliate for the lack of resolution in the near wall region, leading to poor prediction of the friction for very coarse mesh (see Fig 4 obtained with the Localized Dynamic model). Surprisingly, as far as wall friction is concerned, Smagorinsky model seems to be the most accurate model on poor grids, for bad reasons however. Nevertheless, sticking to reasonable resolutions, a clear improvement occurs when more elaborate models such as the Mixed Scale model,the Dynamic model, the Dynamic one-equation model, the Localized Dynamic model or the WALE model are used.

Along the chord of the airfoil, the grid is refined in crucial regions (leading edge, trailing edge, transitional regions, regions of separation) and coarsened in between. The ability of SGS models to restore a coherent dynamics of mean and fluctuating quantities for a flow encountering strong grid variations (i.e. variations of the cut-off length scale) is thus of importance for the airfoil simulation. University of Karlsruhe evaluated the performance of the dynamic SGS model on stretched and distorted grids. The Werner and Wengle wall function [18] provides the near-wall treatment and the dynamic model the SGS stresses. DNS of Moser *et al.* [19] is used as reference data. A $222 \times 38 \times 66$ node grid is used and is $8\pi \times 2 \times \pi$ units in size. Periodic boundary conditions are applied in the streamwise and spanwise directions. Wall normal and spanwise cell sizes are constant at $\Delta y^+ = 30$ and $\Delta z^+ = 60$, respectively. In the streamwise direction, the grid is divided into five different regions. Regions AB and EF contain cells of streamwise size $\Delta x^+ = 30$. The cells in region CD are also of uniform size with $\Delta x^+ = 220$. The cells in regions BC and DE are either stretched or compressed in the streamwise direction at a rate of 5%. Regions AB and EF are π long, while regions BC, CD and DE are each approximately 2π in length.

Figure 5 shows some results from the computation undertaken with the 5% stretched/compressed grid. It can be seen that the grid stretching/compressing has an impact on the time-averaged statistical data. The mean velocity profile appears to be least affected. Variation at the channel centerline is in the order of 1%, increasing to approximately 3% at $y^+ = 63$. The differences between this computation and the computation with a constant $\Delta x^+ = 220$ are of similar magnitudes. The increase and decrease of the streamwise stresses when increasing/decreasing Δx agree with the results in Figs. 4 and 8. A large/small Δx gives large/small streamwise stresses. It should be noted that the evolution of the wall-normal and spanwise stresses is the reverse [20]. This is also seen in Figs. 4 and 8.

The results presented in Fig. 5 show that the stretching in the streamwise direction of a grid in a boundary layer should be kept at a minimum. This finding is a confirmation of earlier findings of Surrey [21] when doing LES of the flow around turbine blades. The results obtained above should be kept in mind when generating the grid for the airfoil.

For their LES of the flow around the cube Fluent have used an adapted mesh, stretched in the cross-stream direction and locally refined around the cube. The streamlines obtained with the WALE model have been plotted in Fig. 6 and a brief comparison of the results in Fig. 7. The overall good results obtained prove that the two models used (Smagorinsky and WALE) behave well on this complicate grid.

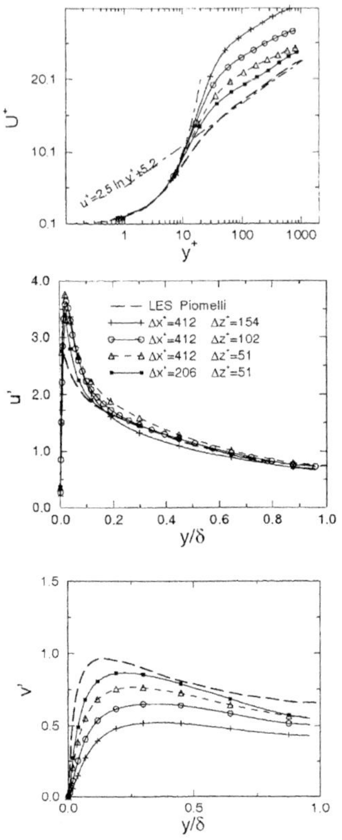

Figure 4 Cerfacs: $Re_\tau = 1050$. Profiles of streamwise velocity and of root-mean-square velocity fluctuations in simulations with LDYN. Grid spacings in the wall normal direction are $\Delta y^+ = 2 - 86$.

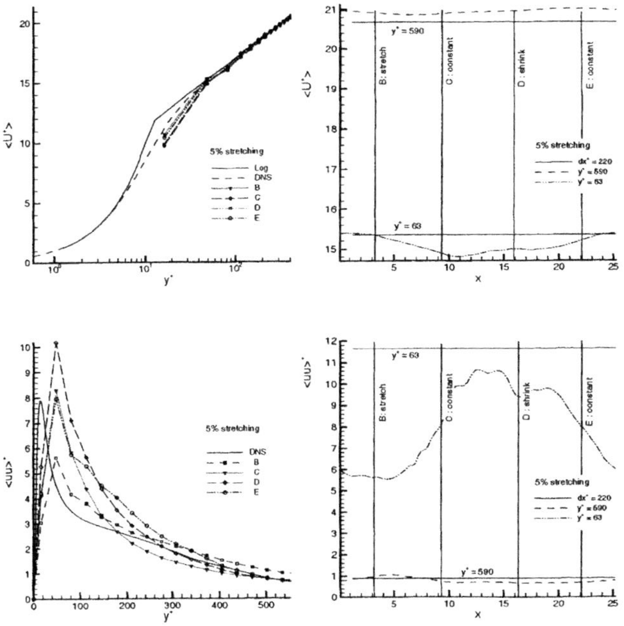

Figure 5 Karlsruhe: Channel flow, $Re_\tau = 590$. Velocity profile and streamwise normal stresses. AB and EF: $\Delta x^+ = 30$; CD: $\Delta x^+ = 220$; BC and DE: Δx^+ stretched/compressed by 5%. Solid lines in the figures to the right show predictions with constant $\Delta x^+ = 220$. Results from Ref. [20].

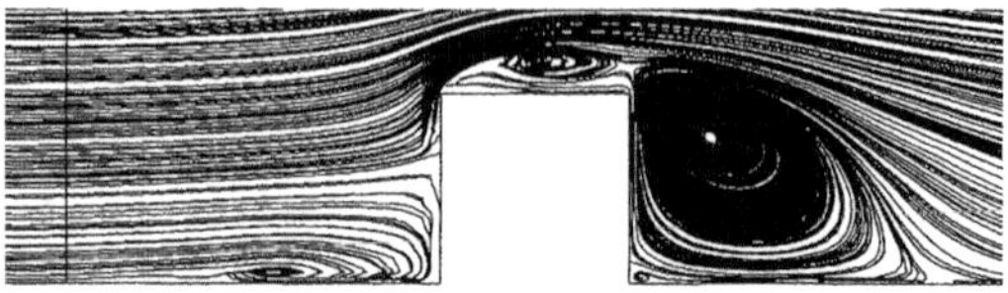

Figure 6 Fluent: Flow around wall-mounted cube. Mean flow streamlines at the centerline (WALE).

Wall Damping. The correct prediction of the wall friction seems to be crucially influenced by the asymptotic behaviour of the subgrid scale model at the wall and by the resolution in all directions. To that respect and for a fixed reasonable resolution, the basic models such as the Smagorinsky and the Structure Function models reveal an incorrect behaviour at the wall, leading to an incorrect prediction of the skin friction. This situation gets better when these models are used in conjunction with wall damping functions which reduces eddy-viscosity close to the wall but the sensitivity of these models to large scales and mean strain makes them produce worse result than selective models.

Numerics. The QUICK scheme, third-order upwind scheme with dissipative properties, has been compared to the second-order centered scheme in the cube test-case. While good agreement with experiment is obtained in the simulation with the central-differencing scheme, QUICK produces poor results and cannot be recommended for use in LES. This is generally the case for any dissipative scheme because numerical dissipation may significantly modify the flow dynamics. Nevertheless, some controlled artificial dissipation may reveal necessary to limit numerical oscillations due to central schemes, particularly at high Reynolds number and on poor grid, see Chapter III.

Along with this conclusion, CERFACS has obtained more accurate results with higher order schemes than with a second-order method. In view of the modest quality of the grids allowed at high Reynolds number by present computer resources, use of higher order schemes can be recommended to achieve better accuracy.

Reynolds number regime. A large range of Reynolds number has been explored. Larger Reynolds number cases, usually performed with coarser grids (in terms of wall units), are more demanding for the models and reveal more clearly their defaults. For example, Chalmers compared their 1-equation Dy-

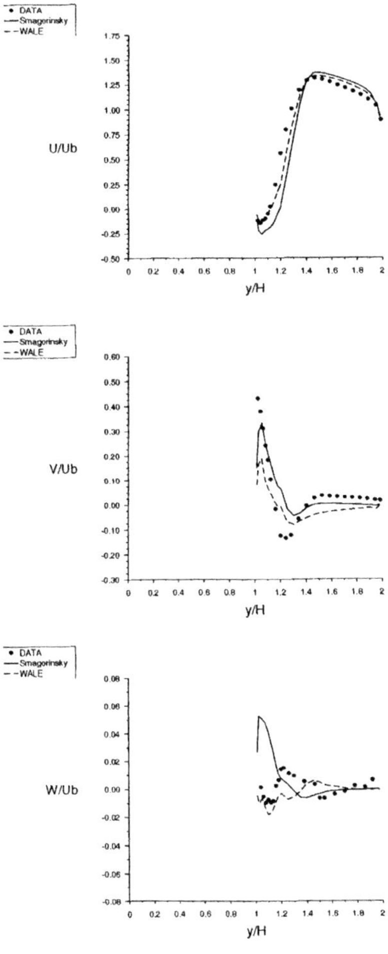

Figure 7 Fluent: Flow around wall-mounted cube. Comparison between predictions and experiment at x/H = 1 (downstream face) and z/H = 0 (symmetry plane)

namic model to the Smagorinsky model in channel flow at $Re_\tau = 395$ and $Re_\tau = 1050$. One important conclusion is that the elaborate models prove to be more robust over a large range of Reynolds number.

The airfoil flow undergoes transition to turbulence both on the suction and pressure sides. Tests of the ability of subgrid scale models to deal with this fundamental change of regime have once again shown that basic models are to be rejected, whereas the Dynamic type model, as well as the filtered structure function and WALE models can be retained.

Wall Function. The Werner and Wengle wall functions [18] was used by several partners and in each of the test-cases. In the channel flow, the use of this wall function with a variety of subgrid-scale models have shown that models hierarchy in term of quality is the same as in wall-resolved simulations (see for example Fig. 8, where the Werner and Wengle wall function has been used together with the WALE model). Forced transition have also been obtained with wall-function. If application of the Werner and Wengle in the channel flow appears quite "natural", its validity in a complex flow including separations is more questionable (this model applies locally and instantaneously a fit of the *law of the wall*, only valid on average in fully developed wall turbulence). Nevertheless, this wall function was also used in simulations of the hill flow and flow over the cube. In such complex flows, it is difficult to make distinction between effects of inappropriate resolution and effects of inappropriate modelling, but at least the approach was found robust and acceptable with results of better quality than those obtained using RANS modeling.

1.5 Conclusion

Task 1 have resulted in a selection of robust subgrid-scale models for large-eddy simulations. These models were shared between partners: UMIST makes use of the Mixed Scale model developed at ONERA, Fluent has implemented and validated the WALE model developed by CERFACS.

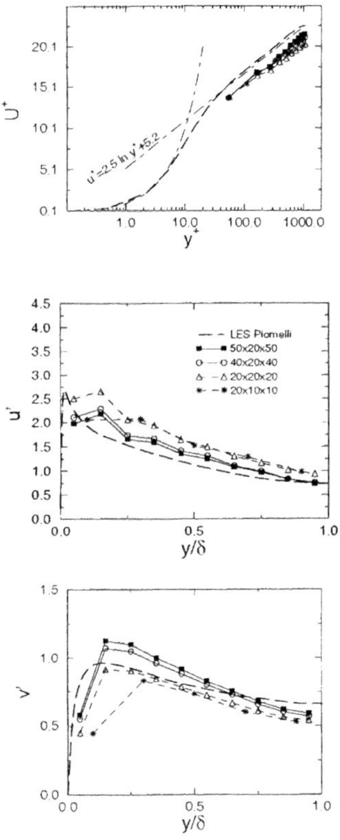

Figure 8 Cerfacs: $Re_\tau = 1050$. Profiles of streamwise velocity and of root-mean-square velocity fluctuations in simulations with the Werner and Wengle wall function and WALE. Resolutions are: $\Delta x^+ = 164, \Delta y^+ = 105, \Delta z^+ = 32$ ($50 \times 20 \times 50$), $\Delta x^+ = 206, \Delta y^+ = 105, \Delta z^+ = 41$ ($40 \times 20 \times 40$), $\Delta x^+ = 412, \Delta y^+ = 105, \Delta z^+ = 82$ ($20 \times 20 \times 20$), $\Delta x^+ = 412, \Delta y^+ = 210, \Delta z^+ = 164$ ($20 \times 10 \times 10$).

2 Task 2: Near-wall models

Task Leader: QMW
M.A. Leschziner and L. Temmerman

2.1 Introduction

The need to resolve the near-wall layer in high-Reynolds-number flows, in which the wall has a strong viscous effect on the behaviour of gross flow features away from the wall, presents a major challenge to LES. This is a major problem in separated airfoil flows in which the separation process hinges to the response of the boundary layer to the adverse pressure gradient on the suction side that arises at high incidence. The difficulty LES faces here is the extremely small thickness of the layer in which viscosity interacts with turbulence and the very fine near-wall grid that is required for its proper resolution. Because LES does not tolerate high cell aspect ratios and more than weak grid-line compression or expansion, grid-refining normal to wall needs to be accompanied by refinement in the streamwise and spanwise directions. This constraint is virtually impossible to meet in practice because of the exorbitant computing costs that would be incurred.

A promising, economically tolerable alternative is to adopt an approximate non-LES treatment to bridge the semi-viscous near-wall region. Thus, the LES procedure is applied in the outer flow region only and excludes a layer, typically, only 1% of the overall scale of the flow. The near-wall approximation can be based on an assumed velocity distribution linking the velocity at the wall-nearest computational node to the wall or on a Reynolds-averaged Navier-Stokes (RANS) methodology which is based on the application of a low Reynolds-number turbulence model. In both cases, the near-wall layer needs to be appropriately coupled to the LES domain, which is not straightforward with the RANS approximation. An additional issue that needs to be addressed in the context of near-wall modelling is the sensitivity of the subgrid-scale model, used in the LES region, to the proximity of the wall. Specifically, there is an established asymptotic behaviour of the subgrid-scale viscosity as the wall is approached, and different subgrid-scale models return this behaviour differently well. The predictive near-wall characteristics of different models need therefore to be investigated and quantified.

The above issues are addressed in the present section by reference to two types of flow: fully developed channel flows at bulk Reynolds number of up to 20000, and a separated channel flow with a streamwise-periodic arrangement of hill-shaped constrictions on the lower wall at a bulk Reynolds number of about 21000, see Fig. 9. The channel flow, apart from being characterised by a single strain component and dominance of a single shear stress, has the merit of being well described by extensive DNS data which permit model performance to be reliably assessed. In particular, satisfaction of the log-law of the wall and the correct prediction of the intense near-wall anisotropy can

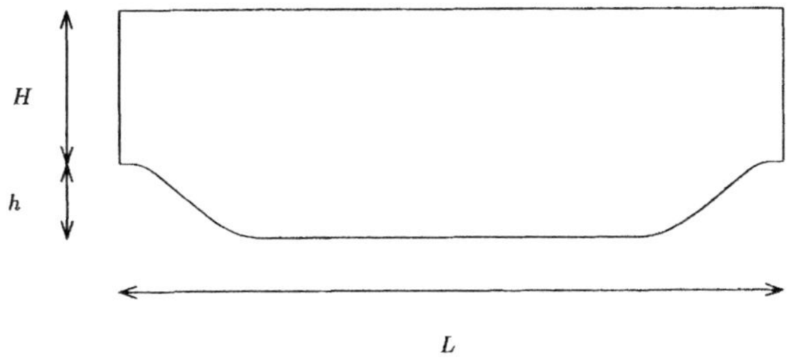

Figure 9 Hill-flow configuration.

be investigated in some detail, and the effect of different near- wall treatment and the influence of near-wall resolution can be examined with a high level of confidence. The hill flow consists, nominally, of an infinite sequence of identical hills, so that the flow can be treated as periodic between consecutive hill crests. The flow separates from the hill's leeward side, reattaches along the long horizontal wall between hills and then partly recovers before being reaccelerated towards the next hill crest. Because of its complexity and the presence of separation, this flow is highly pertinent to separated airfoil flow, in terms of both geometry and flow features.

The section summarises the contributions made by four collaborating partners – CERFACS, Chalmers University, University of Karlsruhe and UMIST/QMW – towards the objective of understanding the role of subgrid-scale models and the advantages and disadvantages of alternative RANS-type near-wall approximations designed to bridge the viscous sublayer at high Reynolds numbers. Efforts are reported in three parts. First, an overview is given in Sub-Section 2.2, in terms of the cases examined and approaches investigated. Then, in Sub-Sections 2.3–2.6, the work is presented in greater detail in partner-related sections. Finally, Sub-Section 2.7 consolidates conclusions and recommendations which fed into subsequent studies of the "Aerospatiale A" airfoil.

2.2 Overview of Research

Investigations reported herein focus on two flows, a fully-developed channel flow at Reynolds numbers ranging between 5000 and 20000, based on mean velocity and channel height, and a *hill* flow formed by channel with streamwise periodic constrictions created by *hills* on the lower wall. The latter flow has a Reynolds number of 21000 and features a large recirculation region between consecutive hills.

All partners investigated the channel flow, some extensively with different grid densities, SGS models and various near-wall approximation methodologies. As noted already, the advantages of channel flow as a test case is its relative simplicity, the availability of reliable DNS data and the fact that statistical properties, obtained by averaging over the two homogeneous directions as well as time, converge rapidly. Hence, this case is economical to compute, thus allowing a wide range of parametric studies to be undertaken.

CERFACS undertook the most extensive investigation of channel flow among the partners, examining some 30 combinations of SGS models over a range of Reynolds numbers, either wall-resolved or in conjunction with a particular near-wall approximation by Werner and Wengle. Particular emphasis was put in these studies on the asymptotic near-wall behaviour of the SGS viscosity, which was shown to be an important issue in wall-resolved simulations. An especially influential aspect of the study, affecting the efforts of other partners, was demonstration of some advantageous properties of two $wall - adapted$ SGS models, referred to as WALE and WAFSF. The former was subsequently adopted by UMIST/QMW for the majority of their simulations, and also investigated by University of Karlsruhe.

Chalmers' major contribution to this task was an examination of hybrid LES/RANS formulations in which the near-wall region is approximated by way of a low-Reynolds-number two-equation model of the type used in conventional RANS computations. With this approach, which is more elaborate and potentially superior one based on wall-function approximations, particular problems arise in relation to the interface between the RANS and LES regions, and this issue has been on focal point of Chalmers' efforts. Studies were performed for both the channel and the hill flow. In the former, the hybrid method was not found to yield the expected advantages, especially when the RANS near-wall layer is relatively thick. A more favourable behaviour was observed in the hill flow, argued to be a consequence of the greater importance of cross-interface convection, relative to diffusive transport.

Like Chalmers, Karlsruhe also investigated both the channel and hill flows, devoting special attention and effort to the latter. For channel flow, Karlsruhe mainly investigated the sensitivity of predicted flow properties to the thickness of the near-wall layer in which the Werner-Wengle wall-function approach was applied, demonstrating a progressive decline in accuracy with increasing near-wall layer thickness. Karlsruhe made an exceptionally large contribution to the hill-flow investigation, through costly preliminary simulations leading to the definition of the final hill geometry, the provision of grids for this geometry and the generation of reference data from a highly-resolved and extremely resource-intensive simulation over a very fine grid. These data – supplemented by additional data generated later by a similar simulation undertaken by UMIST/QMW – played a central role in assessing the accuracy of approximate solutions performed over coarser grids with near-wall approximations. Karlsruhe also investigated two quite different types of wall-function approximations for the hill flow, one being the conventional Werner-Wengle method and the other a new *back-flow wall function* designed specifically for reverse-flow regions. Both were tested for two grids, and found to yield similar levels of accuracy.

UMIST/QMW's studies paralleled, in part, those of Karlsruhe, but involved more extensive investigations of both channel and hill flow, undertaken with a wider range of SGS models and near-wall approximations. Like Karlsruhe, UMIST/QMW also performed a highly-resolved fine-grid simulation, intended to strengthen the credibility of the reference data generated by Karlsruhe, and laying the foundation for *a-priori* studies of wall functions as well as investigations of realisability issues and of the budgets of the Reynolds stresses. For the channel flow, some 17 simulations were undertaken with 8 SGS models and 4 wall-function variants. As done previously by CERFACS, particular attention was paid to the asymptotic near-wall behaviour. This study led to the subsequent preference for the WALE SGS model in combination with the Werner-Wengle wall function. For the hill flow, extensive studies were performed over three parametric *axes* – namely, grid density, SGS modelling and near-wall approximation. Here some 20 simulations were performed to reveal a strong variability of the solutions and sensitivity to all three parametric variables. Overall, the combination of the WALE model and the Werner-Wengle wall-function was found to be the best, provided grid resolution was adequate especially in the vicinity of the separation point downstream of the hill crest.

Below partner-related efforts are presented.

2.3 Chalmers

Author: L. Davidson

Chalmers' efforts were directed, principally, towards examining methods which couple LES with RANS-type models for the near-wall region. In essence, this involves the use of a turbulence model, applied in unsteady mode, to bridge the gap between the wall and the inner LES region. The advantage of this approach lies in the fact that low-Reynolds-number turbulence models tend to give (at least in steady flow) a satisfactory detailed representation of the viscous sub-layer, even with extremely high-aspect-ratio grids. Thus, their use with LES circumvents the need for grid refinement in all 3 directions that would be necessary for wall-resolved simulations.

Chalmers' approach couples a two-equation low-Reynolds-number (LRN) $k-\omega$ model in the near-wall region with a one-equation k_{SGS} model in the core LES region, as shown in Fig. 10. The momentum equations are solved throughout the computational domain. In the RANS region, the turbulent viscosity from the $k-\omega$ model is used, while in the LES the SGS model provides the viscosity.

For simplicity, the matching line is defined, in the current implementation, at a pre-selected grid line. This approach is similar to the DES (Detached Eddy Simulation) which couples a one-equation model to the LES core region in separated flow [22–25]

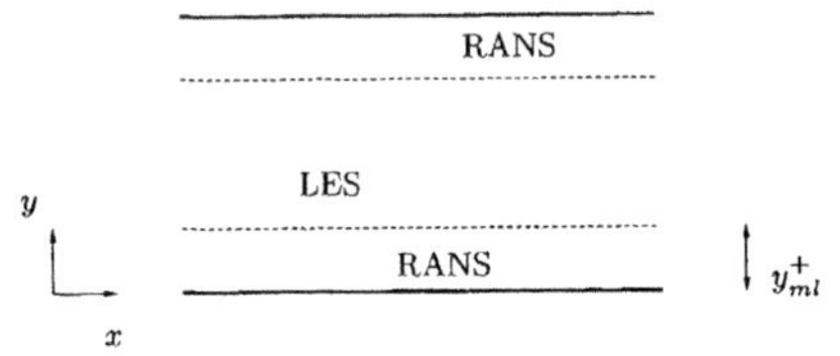

Figure 10 The near-wall RANS region and the outer LES region.

Table 3 Channel flow, $Re_\tau = 1050$, $x_{max} = 4\pi$, $\Delta x^+ = 412$. Size of the computational domain and position of the matching line (ml) between the LES and RANS regions. The j_{match} value represents number of cells in the RANS region. Note that in Case 4 only LES is used.

Case	z_{max}	y_{ml}/δ	j_{match}	y^+_{ml}	Δz^+
1	π	0.023	4	25	104
2	π	0.057	8	60	104
3	2π	0.057	8	60	208
4	π	0	0	0	104

The Hybrid LES–RANS model. The RANS near-wall model is the a $k-\omega$ model of Peng et al [26]. In the LES region, the one-equation SGS model by Yoshizawa [27] is used. The latter reads

$$\begin{aligned} &\frac{\partial k_{sgs}}{\partial t} + \frac{\partial}{\partial x_j}(\bar{u}_j k_{sgs}) = \frac{\partial}{\partial x_j}\left[(\nu + \nu_{sgs})\frac{\partial k_{sgs}}{\partial x_j}\right] + P_{k_{sgs}} - C_\varepsilon \frac{k_{sgs}^{3/2}}{\Delta} \\ &P_{k_{sgs}} = 2\nu_{sgs}\bar{S}_{ij}\bar{S}_{ij},\ \nu_{sgs} = C_k \Delta k_{sgs}^{1/2},\ \Delta = \min\{\Delta_\xi, \Delta_\eta, \Delta_\zeta\} \\ &\qquad C_k = 0.07,\ C_\varepsilon = 1.05 \end{aligned} \tag{1}$$

where ξ, η and ζ denote the three coordinate directions defined by a general curvilinear grid. The coefficients have been assigned slightly different values to those above [28]. Contrary to the standard Yoshizawa model, in which $\Delta = (\Delta_\xi \Delta_\eta \Delta_\zeta)^{1/3}$, Chalmers use the smallest cell side, which was found to considerably improve the predictions.

The matching line at a wall (here the lower wall) is located at y_{ml}, see Fig. 10. If j_{match} is taken to denote the cell below the matching line y_{ml}, the following matching conditions are used:

$$
\begin{aligned}
&j = j_{match}: \ \frac{\partial k}{\partial y} = \frac{\partial \omega}{\partial y} = 0 \\
&j = j_{match} + 1: \ \nu_{sgs, j_{match}+1} = \nu_{t, j_{match}} \Rightarrow k_{sgs}^{-} = \left(\frac{\nu_t}{C_k \Delta}\right)^2_{j_{match}}
\end{aligned}
\tag{2}
$$

where ν_t denotes the RANS turbulent viscosity, and k_{sgs}^{-} is the SGS kinetic energy which is transported by convection and diffusive transport across the matching line from the RANS region to the LES region. The above matching at the interface imply that k_{sgs} in the LES region is determined from the RANS region, but that k and ω in the RANS region are not influenced by k_{sgs} in the LES region.

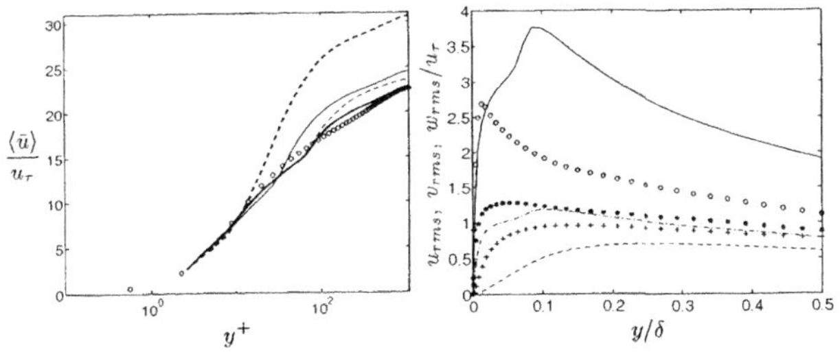

(a) Thin solid line: Case 1; thick solid line: Case 2; thin dashed line: Case 3; thick dashed line: Case 4.

(b) Case 2. Resolved, rms fluctuations.

Figure 11 Chalmers: Channel flow. Markers: LES by Piomelli [29].

The above methodology has been implemented and tested for both channel flow and the separated hill flow. Representative results are given in Figs. 11–14, the first two relating to channel flow and the last two to the hill flow.

In Fig. 11a, the predicted $\langle \bar{u} \rangle$-velocities ($\langle . \rangle$ denotes averaging over x, z and t) are shown for fairly coarse meshes, see Table 3, all generated with $32 \times 64 \times 32$ (x, y, z) nodes. For Case 2, in which the matching line was located at $y^+ = 60$, and $\Delta z^+ = 104$, agreement is seen to be reasonably good. However, a small kink is visible in the $\langle \bar{u} \rangle$ near the location of the matching line. When the resolution is coarsened in the spanwise direction (Case 3), agreement relative to the benchmark LES deteriorates. It is important to point out that the grid for Case 2 & 3 is much coarser than that required for a wall-resolved LES. Consequently, when only LES is used (Case 4), without any wall model, the results are much worse than those obtained with the LES–RANS model. Figure 11b presents the predicted resolved, rms fluctuations. Agreement with the benchmark LES is clearly not close. The predictions are typical of an under-resolved LES: thus, the streamwise fluctuations are too large, and the wall-normal and spanwise fluctuations are too small. It should be noted that even in the RANS region the resolved stresses are large, which reinforces the conclusion that insufficient resolution is an important factor in the disagreement observed.

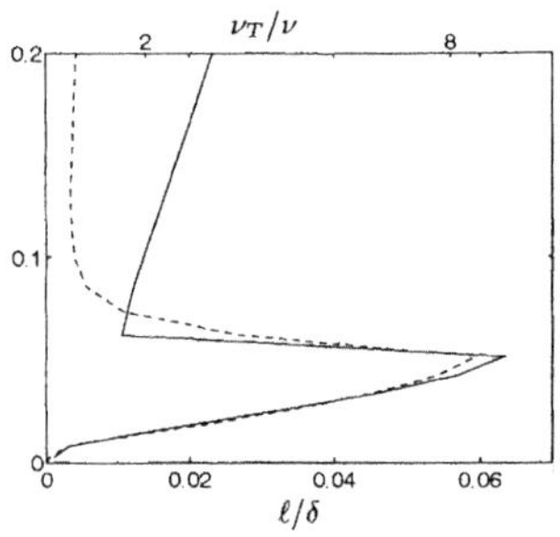

Figure 12 Chalmers: Channel flow, Case 2. Turbulent/SGS viscosity and length scales. For $y \leq y_{ml}, \nu_T = \nu_t$, $k_T = k$, otherwise $\nu_T = \nu_{sgs}$, $k_T = k_{sgs}$. Dashed line: turbulent/SGS viscosity. Solid line: turbulent/SGS length scales, $\ell = \langle \nu_T \rangle / (C_k \langle k_T \rangle^{1/2}$.

For the hill flow, a $104 \times 64 \times 32$ (x, y, z) grid was used. The matching line along the lower wall was fixed to coincide with grid line 13, so that the location of the matching line is $n_{ml}/h = 0.1 \pm 0.02$, which corresponds to $4 \leq n_{ml}^+ \leq 87$. Along the upper wall no RANS region is used, but the LES region extends up to the wall. This treatment is inaccurate, but does not impinge on the validity of the observations derived for the lower wall. The cell size Δx^+ and Δz^+ is kept below 25 and 100, respectively, except at the hill crest, where they reach values close to 100 and 300, respectively. For the nodes adjacent to the lower wall, $\langle y^+ \rangle \lesssim 2$. The predictions are compared with a benchmark LES reported in Refs. [20, 30] performed with a grid of around 5 million cells. In Fig. 13 the predicted $\langle \bar{u} \rangle$ velocities ($\langle . \rangle$ denotes averaging over z and t) and resolved, rms fluctuations are compared with benchmark LES. Agreement is good. Minor kinks are visible in the $\langle \bar{u} \rangle$ profiles close to the location of the matching line. These are much smaller than those observed in the channel flow (cf. Fig. 11). The reason may be that in the hill flow the convective transport across the matching line is considerable, which has a smoothening effect.

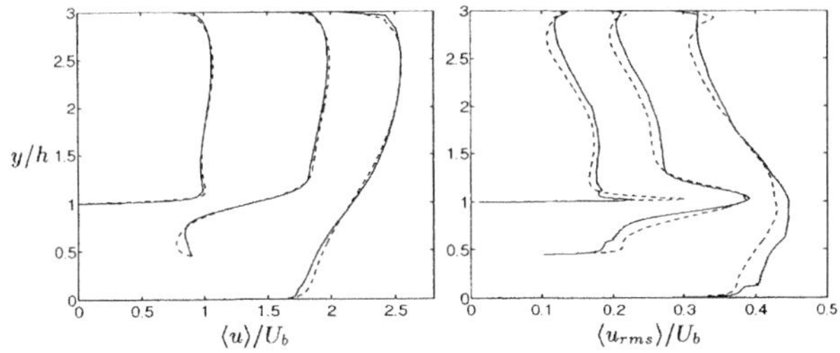

Figure 13 Chalmers: Hill flow. Profiles at $x = 0.005h, x = h, x = 6h$.

Important insight into the compatibility between the RANS and LES regions is gained from examining the turbulent and SGS viscosity distributions. Such distributions are shown in Figs. 12 and 14 for the channel and hill flows, respectively. It can be seen that, in the RANS region, the turbulent viscosities are large – much larger than the level normally found in LES regions. Near the matching line ν_{sgs} in the LES region drops sharply, down to typical SGS values of the order $2 \leq \nu_{sgs}/\nu \leq 4$. The maximum value of ν_t (Fig. 12) in the RANS region, which occurs near the matching line, is $\nu_{t,max}/\nu \simeq 9$. When 1D, steady RANS is used to compute the channel flow using the same grid and the same $k - \omega$ model, the value of ν_t at the same location is approximately twice as large. One reason why the RANS calculation gives larger value of ν_t than the hybrid LES-RANS, is that in the former case less modelled turbulent kinetic energy is diffused to the core region. Another reason is that in the hybrid LES-RANS computations, a substantial part of the turbulence in the RANS region is represented by resolved turbulence, see Fig. 11b. Figs. 12 also presents the modelled turbulent length scale. The modelled turbulent/SGS viscosity can be defined as $\nu_T \propto \mathcal{U}\ell$, where $\mathcal{U}$ and ℓ represent a turbulent (SGS) velocity scale and length scale in the RANS (LES) region, respectively. In the LES region, the SGS length scale is $\ell = \Delta \equiv \nu_{sgs}/(C_k k_{sgs}^{1/2})$, see Eq. 1. In order to compare the turbulent and the SGS length scales, the turbulent length scale is defined in the same way, so that $\nu_T = C_k k_T^{1/2} \ell$ in both the RANS and the LES region. As can be seen from Figs. 12, the turbulent/SGS length scale decreases steeply from the RANS region to the LES region. Thus, the turbulent/SGS kinetic energy must increase to balance the decrease in the length-scale in order to satisfy $\nu_{sgs} = \nu_t$, see Eq. 1. A large peak in k_{sgs} is created just outside the matching line. Of course, the modelled SGS kinetic energy k_{sgs} in this region is not physical. However, this is of no consequence, since the modelled quantity which enters the momentum equation is the turbulent/SGS viscosity ν_T, not k_T. The advantage of the condition $\nu_{sgs} = \nu_t$ across the matching line is that the gradient in the turbulent/SGS viscosity across the matching line is reduced, because the gradient of k_T is smoothed out by convection/diffusion of k_T.

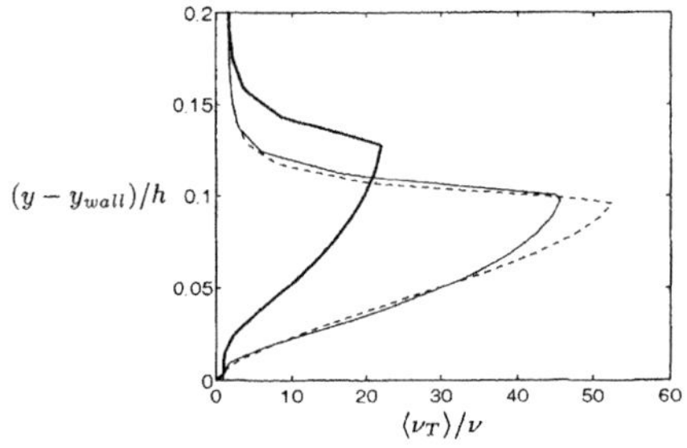

Figure 14 Chalmers: Hill flow. Turbulent/SGS viscosity. Dashed line: $x/h = 0.005$; thick solid line: $x/h = r$; thin solid line: $x/h = 6$.

2.4 CERFACS

Authors: P. Moinat, F. Ducros

CERFACS principal contribution was to investigate the role and performance of a range of SGS models in channel flow, with emphasis on the near-wall behaviour of the models. In addition, CERFACS examined hybrid approaches, which couple wall-function representations of the near-wall layer to the outer LES region. This methodology is a simpler alternative to the LES-RANS approach investigated by Chalmers. Table 4 provides an overview of the parametric studies undertaken. Aspects investigated included numerical resolution, SGS modelling, wall function approximation, and Reynolds-number dependence.

An important issue underpinning the rationale of CERFACS' investigations is the asymptotic behaviour of SGS models close to the wall. This behaviour influences the overall structure of the simulated flow, and is especially pertinent to wall-resolving simulations which do not adopt an approximate near-wall model. Taylor-series expansions show that the eddy-viscosity ν_t should decrease, close to the wall, as a function of the distance y to the wall in accord with $\mathcal{O}(y^3)$. This is not satisfied by the classical Smagorinsky model [31] in which the viscosity scales with $(\Delta_{eq} C_S)^2$, *ie* like $\mathcal{O}(1)$. A possible correction is the application of a near-wall damping function, leading to a scaling which accords with y^2 or y^3. As seen from Table 4, the introduction of damping functions influences the simulated flow: the prediction of the mean velocity profile and of the wall shear stress (represented by the wall-friction Reynolds number) is somewhat improved. However, the effect is not large and certainly of secondary importance in comparison, for example, to the effect of a modification in the constant C_S. CERFACS thus tested 6 other eddy-viscosity models in simulations of the turbulent channel flow at Reynolds numbers 180 and 1050 (based on friction velocity and channel half width): the Smagorinsky model (SMAG) (with van Driest wall damping), the dynamic Smagorinsky model (DYNA), the localized dynamic Smagorinsky model (LDYN), the WALE model, the filtered structure function model (FSF) and the wall-adapted filtered structure function model(WAFSF). Except for the Smagorinsky model, all formulations are selective, *ie* they are not sensitive to the mean flow but to the resolved turbulence.

The investigation revealed that dynamic models give, without specific practices, the correct y^3 behaviour for ν_t. Accurate results can be obtained with these models, provided the resolution is adequate. The WALE model was especially designed to produce the correct wall behaviour, but also has other desirable qualities: it is local and can handle transition. It has thus been promoted by CERFACS and tested by other partners. In fact, UMIST/QMW has favoured this model and adopted it for the majority of its simulations. The results (mean velocity, shear stress and velocity fluctuations) obtained with WALE are close to the ones obtained with dynamic models. The FSF model yields an eddy-viscosity which varies as $\mathcal{O}(y)$ near the wall, and is thus excessively dissipative close to walls. Much better results are obtained with the WAFSF model, although it was found less accurate than dynamic or WALE models.

The discussion of eddy-viscosity scaling in near-wall regions is only meaningful if the wall layer is resolved. But even on coarse grids ($\Delta_x^+ = 200, \Delta_y^+ = 4, \Delta_z^+ = 50$) which obviously do not resolve correctly the near-wall structure, is was observed that models yielding the correct scaling give much better results. It should be stressed, however, that none of the model is able to correctly represent the effect of the near-wall turbulence when the near-wall layer is poorly resolved. For all models, under-resolved simulations yield low wall stress, excessive streamwise fluctuations and too low cross-stream and spanwise fluctuations. This is in line with observations made by other partners, and has already been noted in relation to Chalmers' results. Sometimes, compensation of errors is observed: under-resolved simulations with an over-dissipative model can produce surprisingly good (but fortuitous) results.

Because wall-resolved LES is not practically possible for high-Reynolds number flows, CERFACS extended the above investigation to hybrid formulations which couple LES with approximate near-wall models, using some of the SGS models investigated above in the context of wall-resolved simulations. As a precursor, an CERFACS investigated some basic issues of hybrid formulations by combining two different SGS models in different subdomains of a pure LES. In one such combination, the SMAG model was applied close to the wall, in the poorly resolution region, with the LDYN model further from the wall where resolution is more adequate. The expectation was that the combination would give good wall-friction characteristics with favourable flow dynamics. The hybrid combination was constructed as a weighted linear blending: $HYB = p \times SMAG + (1 - p) \times LDYN$. The weighting parameter p is set to 1 at the wall and 0 for $y^+ > Y_S$. Three values of Y_S were investigated: 200, 50 and 25. Simulations were performed on a $\Delta x^+ \times \Delta y^+ \times \Delta z^+ = 412 \times (2 - 86) \times 154$ grid. The results turned out to be disappointing, simply yielding a solution intermediate between those for the SMAG and LDYN models applied separately. Thus, for $Y_S = 200$, the behaviour is close to that obtained with the SMAG model, while for $Y_S = 25$, it is close to that obtained with the LDYN model (see Table 4 and mid-term report).

As the next step, two wall-function near wall approximations were investigated, one based on the log-law of the wall and the other being the Werner and Wengle (WW) model, the latter based on a power-law approximation to the log-law profile. Some results are shown in Fig. 2.4 for the combination of the WALE model with the WW wall function. Results with 4 grids are included in comparison with an LES bench-mark solution. Overall, the agreement is fairly satisfactory, even on the coarsest grid. As expected, the peaks of velocity fluctuations cannot be properly captured, but the simulations give acceptable estimations of their amplitude.

A comparison has also been undertaken, for one grid ($20 \times 20 \times 20$), between the WW and the log-law implementations. The results were found to be similar, but not identical. In particular, the predicted turbulent Reynolds numbers differs by 6% difference between the two simulations. This is, however, within the range of error associated with the approximation to the logarithmic law by the power law in the Werner and Wengle model.

A final aspect covered here relates to the application of selected subgrid-scale models to the airfoil flow, which is the subject of Chapter III. Thus,

Fig. 16 shows the profile of SGS viscosity arising from the WALE and WAFSF models at one chord position on the suction side of the profile, as evaluated from a wall-resolved simulation detailed in Chapter III. As seen, the requisite wall-damping behaviour is recovered, illustrating that the conclusions drawn for the channel-flow simulations also hold for more demanding realistic applications.

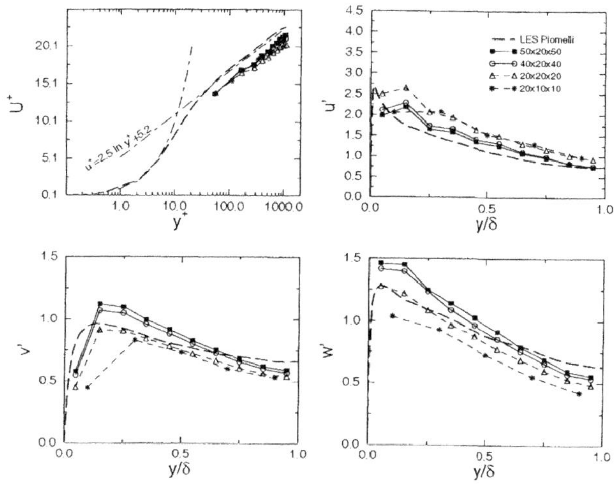

Figure 15 Cerfacs: $Re_\tau = 1050$. Profiles of streamwise velocity and of root-mean-square velocity fluctuations in simulations with the Werner and Wengle wall function and WALE. Resolutions are: $\Delta x^+ = 164, \Delta y^+ = 105, \Delta z^+ = 32$ ($50 \times 20 \times 50$), $\Delta x^+ = 206, \Delta y^+ = 105, \Delta z^+ = 41$ ($40 \times 20 \times 40$), $\Delta x^+ = 412, \Delta y^+ = 105, \Delta z^+ = 82$ ($20 \times 20 \times 20$), $\Delta x^+ = 412, \Delta y^+ = 210, \Delta z^+ = 164$ ($20 \times 10 \times 10$).

2.5 University of Karlsruhe

Authors: C. Mellen, J. Fröhlich, W. Rodi

Karlsruhe's efforts concentrated primarily on evaluating the applicability of wall function models to both channel flow and the recirculating hill flow. This was done in close coordination and collaboration especially with UMIST/QMW, which undertook a study supplementing that of Karlsruhe.

Table 4 $Re_\tau = 1050$. Unless otherwise specified, $C_S = 0.1$ for SMAG, $C_W = 0.1$ for WALE and $\alpha^{(2)} = 0.0043$ for WAFSF. The wall-damping applied to the Smagorinsky model is indicated with the resulting behaviour (y^2 or y^3). This table explains the parametric studies concerning resolution, SGS models, wall functions, and Reynolds numbers.

$\Delta x^+, \Delta y^+, \Delta z^+$	model	Re_τ	U_c/U_b
412, 105, 82	SMAG + WW	1021	1.13
412, 105, 82	LDYN + WW	1127	1.10
412, 105, 82	WALE + WW	1131	1.11
412, 105, 82	WALE + log	1079	1.10
412, 210, 164	SMAG + WW	997	1.16
412, 210, 164	WALE + WW	1099	1.11
206, 105, 41	WALE + WW	1098	1.12
164, 105, 32	WALE + WW	1084	1.12
412, 2-86, 154	DYNA	791	1.078
412, 2-86, 154	LDYN	746	1.076
412, 2-86, 154	WALE	798	1.073
412, 2-86, 154	FSFM	1144	1.158
412, 2-86, 154	WAFSF	735	1.092
412, 2-86, 154	SMAG y^3	1047	1.1
412, 2-86, 154	HYB1 $Y_S = 200$	1015	1.098
412, 2-86, 154	HYB2 $Y_S = 50$	979	1.098
412, 2-86, 154	HYB3 $Y_S = 25$	877	1.088
412, 2-86, 102	LDYN	831	1.073
412, 2-86, 51	LDYN	913	1.073
206, 2-86, 51	LDYN	956	1.095
257, 4-73, 51	SMAG y^3	1054	1.097
257, 4-73, 51	SMAG y^2	1100	1.099
257, 4-73, 51	SMAG y^2 $C_S = 0.18$	1090	1.118
257, 4-73, 51	DYNA	1005	1.086
257, 4-73, 51	LDYN	976	1.087
257, 4-73, 51	WALE	1014	1.084
257, 4-73, 51	WALE - Scotti	971	1.086
257, 4-73, 51	WALE $C_w = 0.3$	953	1.084
257, 4-73, 51	WAFSF	934	1.089
257, 4-73, 51	WAFSF $\alpha^{(2)} = 0.002$	959	1.087
	Expected	1050	1.110

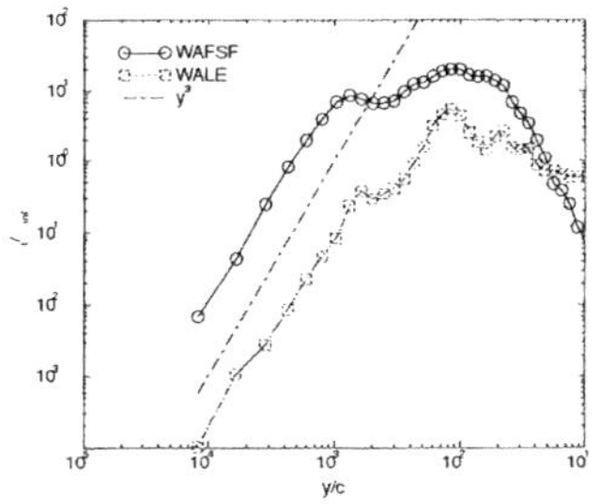

Figure 16 Cerfacs: Spanwise averaged eddy-viscosity profiles at $x/c = 0.7$ on the suction side for a wall resolved simulation around the A airfoil. The right y^3 behaviour is shown.

Particular attention was directed towards the applicability of the Werner-Wengle near-wall model and a new wall law especially formulated for recirculating flow.

For the channel flow, a series of simulations were undertaken with the objective of evaluating the performance of the Werner-Wengle wall model [18] at high Reynolds number and large near-wall cell sizes. For all computations the Reynolds number based on bulk velocity and half channel height was $Re = 10935$ and the dynamic SGS model was employed. The computations used a grid which was uniform in the x and z directions, with the cell size being $\Delta x^+ \sim 60$ and $\Delta z^+ \sim 30$ in all cases. The difference between the computations lay in the cell size normal to the wall. Statistical properties were obtained by averaging in the homogeneous $x - z$ plane for at least $50t_u^+$ (non-dimensional time units based on friction velocity). With the exception of the grid used for LES 4, all grids were uniformly spaced in y. The physical extent of the computational domain was set to be $4\pi \times 2 \times \pi$. Simulations are compared to DNS data of Moser *et al.* [19] at $Re_b = 10935$ DNS. Results were evaluated by assessing the predicted wall-friction Reynolds number $\langle Re_\tau \rangle$, the mean streamwise velocity profile $\langle u \rangle^+$ and the near wall fluctuating quantities $\langle u_i u_j \rangle^+$. Results for the wall-friction Reynolds number are given in Table 5.

As seen, computations gave a Re_τ which agreed reasonably well with the nominal value and with the DNS data. The maximum discrepancy is only 2.9%. Figs. 17 and 18 show $\langle u \rangle^+$ and $\langle uu \rangle^+$. Evidently, the predicted velocity suffers as a result of the relatively coarse grid resolution, but the behaviour in the log-law regions is reasonably well reproduced.

A major effort was expended by Karlsruhe on the separated hill flow. Indeed, Karlsruhe guided the collaborating partners in terms of decisions on the choice of geometry, flow parameters and grids to be used. A sequence of 3

Table 5 Matrix of Karlsruhe channel computations († relates to wall cells; †† relates to interior cells).

Comp	Cell spacing	Calculated Re_τ	% Diff (wrt DNS)
DNS		584.1	
LES 1	$\Delta y^+ \sim 15$	590.7	1.1
LES 2	$\Delta y^+ \sim 31$	601.3	2.9
LES 3	$\Delta y^+ \sim 73$	597.1	2.3
LES 4	$\Delta y^+ \sim 31^\dagger$ to $127^{\dagger\dagger}$	595.4	1.9

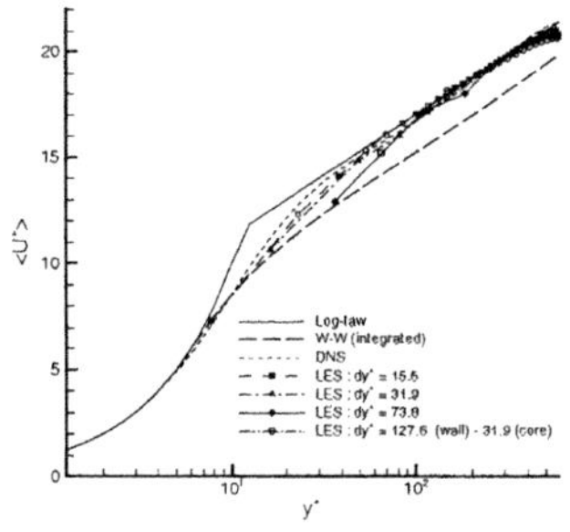

Figure 17 Karlsruhe: $\langle u\rangle^+$ profiles

grids were generated by Karlsruhe and adopted by the other partners. Thus, a major feature of the interaction in this task was that computations for the hill flow were done on common grids. A primary objective of the work was the evaluation of wall models in recirculating flows.

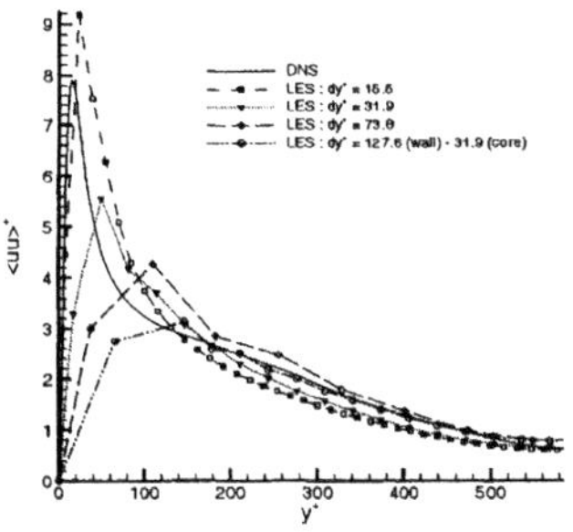

Figure 18 Karlsruhe: $\langle uu \rangle^+$ profiles

One of the more important and time-consuming tasks was the definition of a new geometry which was viable in terms of resource requirements and which contained well-established flow separation and reattachment. This geometry is based upon the periodic hill experimentally investigated by Almeida [10]. Modifications were introduced which lower the cost of the computation while generating substantial variability in the reattachment location of the recirculation region. A comprehensive description of the geometry and the primary flow features are given in [30] and [32]. Essential to the usefulness of this new geometry was the completion of a wall-resolving LES of the flow – a computation also undertaken by QMW/UMIST, albeit with a different SGS model. Taking approximately 60000 CPU hours to complete on an IBM RS 6000 SP (the QMW/UMIST computations required approximately 30000 CPU hours on a 256-processor partition of a Cray T3E), this computation provided a reference against which different wall models could be evaluated. The wall resolving mesh employed was of very high quality. It contained 202 × 130 × 196 points and was constructed so that the grid point closest to the lower wall had a wall-normal distance of less than $y^+ < 1$ nearly everywhere, as shown in Fig. 19. Additionally, 10 or more mesh points were placed below $y^+ = 10$ over most of the wall while the streamwise and spanwise cell sizing was such that $\Delta x^+ < 25$ and $\Delta z^+ < 15$ for nearly the whole of the near-wall region. The dynamic model SGS was used, and the computation was continued until the flow statistics had been averaged for approximately 55 flow through times.

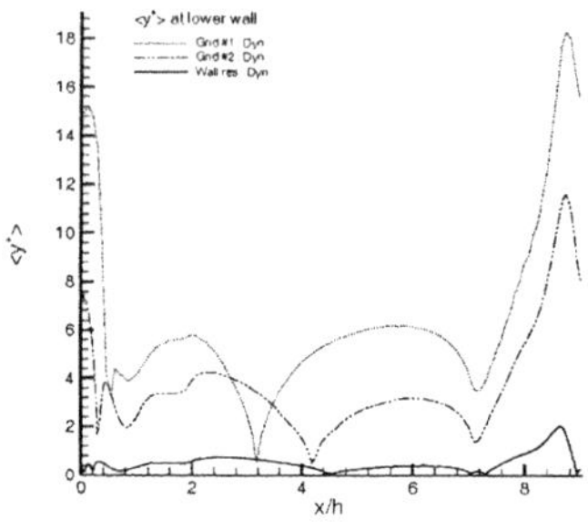

Figure 19 Karlsruhe: y^+ at lower wall

With the above fine-grid simulations providing reference data, two different near-wall models were evaluated on this separated flow test-case. One was the Werner-Wengle model [18]. This model was originally conceived for use with attached flows only. The second near-wall model evaluated was specifically intended for use in regions of recirculating flow [33]. This model assumes that the velocity profile inside a region of recirculation may be represented by

$$U^* = \frac{1}{\kappa}(y^* - ln(y^*) - 1) \tag{3}$$

where $y^* = y/y_N$, $U^* = (U + |U_N|)/U_\tau$, $|U_N|$ is the magnitude of the maximum backflow velocity and y_N is the distance from the wall to the point of maximum backflow. The validity of this wall function was assessed in *a priori* fashion using flow profiles taken from within the recirculation zone of the wall resolving computation. When normalised using the backflow scaling, the computational profiles matched very well with Eq.3, as shown in Fig. 21. Fig. 20 shows, in contrast, that when normalised to correspond to a log-law assumption, i.e. with U_τ and $l_\tau = \nu/U_\tau$, the agreement between these computational profiles and the log-law profile is quite poor. The *a posteriori* assessment of the wall law is contained in Table 6. LES with the new wall

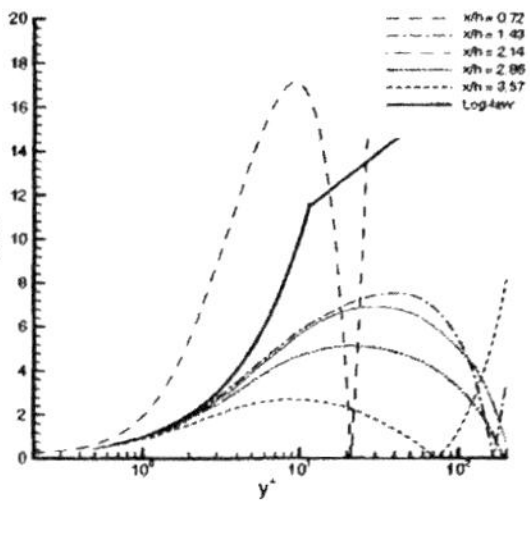

Figure 20 Karlsruhe: Wall-unit scaling

model are reported together with results obtained with the classical wall function of Werner and Wengle. All computations employed the dynamic SGS model. Two different wall-function meshes were used. Their details can be found in [20]. The wall resolving computation is also included there.

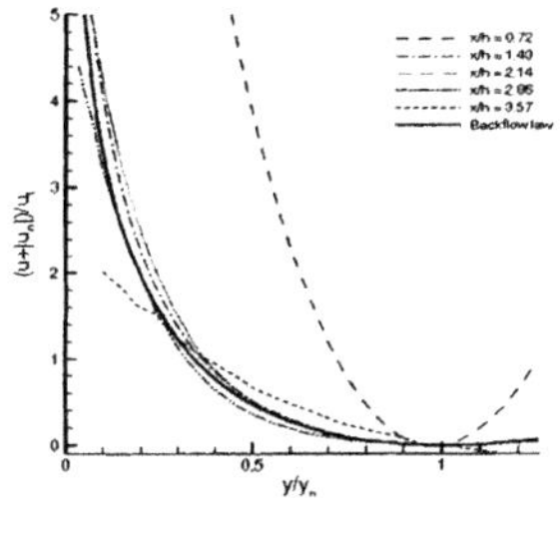

Figure 21 Karlsruhe: 'Backflow' scaling

Fig. 22 shows the shear stress $\langle\tau_w\rangle$ at the lower wall given by computations 2, 3 and 4. For both wall models, $\langle\tau_w\rangle$ on Grid 2 compares favourably to that of the wall resolving computation. Although not shown here, the result on Grid 1 suffers from the relatively poor resolution of this grid.

Moreover it appears that in this case no particular benefit can be gained from application of the backflow function as its prediction of the wall shear stress is not appreciably more reliable than that of the Werner-Wengle law. Though not shown, the results for the other flow quantities of interest (such as $\langle U \rangle$ and $\langle u_i u_j \rangle$) support conclusions similar to those above.

Table 6 Matrix of Karlsruhe hill computations

Comp #	Wall law	SGS model	Grid	Integration time
1	W-W	Dyn	118 × 66 × 96	~ 115 f.t.t
2	W-W	Dyn	182 × 66 × 96	~ 100 f.t.t
3	'Backflow'	Dyn	182 × 66 × 96	~ 100 f.t.t
4	no-slip	Dyn	202 × 130 × 196	~ 55 f.t.t

When assessing the results of these wall function computations, the distance to the wall of the first computational point is of interest, and this is given in Fig. 19. Inside the recirculation zone y_1^+ for both wall function meshes is substantially less than 10. In this range, the Werner-Wengle wall law defaults to the no-slip condition. Strictly then these are not wall function meshes. However, the magnitude of this wall normal spacing is relevant to the conditions encountered in airfoil simulations.

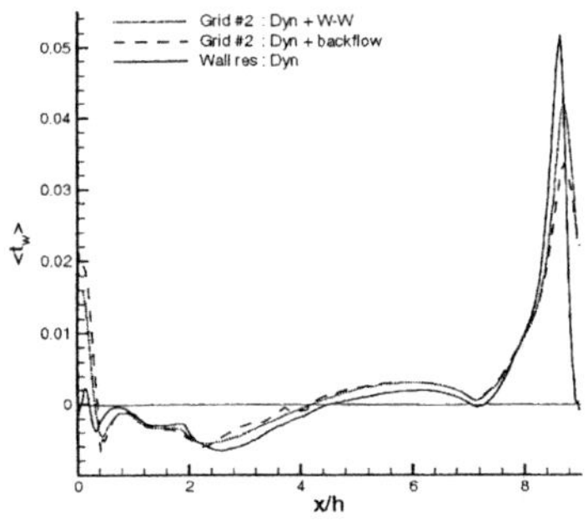

Figure 22 Karlsruhe: τ_w at lower wall

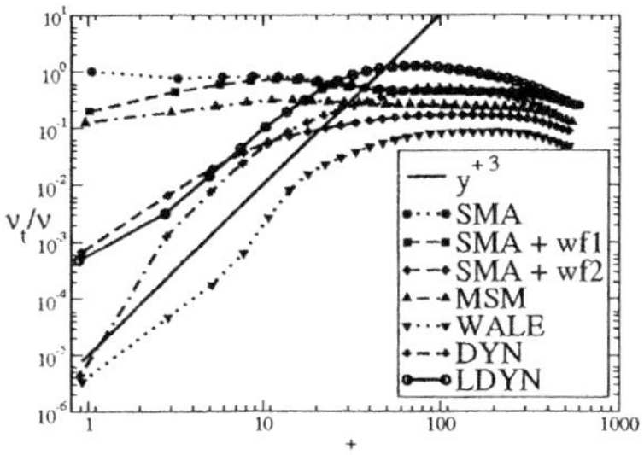

Figure 23 QMW: Distribution of subgrid-scale viscosity $\frac{\langle \nu \rangle}{\nu_t}$ vs y^+

2.6 UMIST/QMW

Authors: L. Temmermann, M. Leschziner

UMIST/QMW's work programme involved simulations for both channel flow and the hill flow, with particular emphasis placed on an investigation of SGS modelling and wall-function performance in both attached and separated conditions. To a considerable extent, UMIST/QMW's efforts paralleled those of Karlsruhe, though a larger ranges of SGS and wall-function practices were examined. Throughout the work, close liaison was maintained with Karlsruhe in efforts to optimise commonality as well as complementarity of practices investigated. In particular, the same meshes were used for the hill simulations.

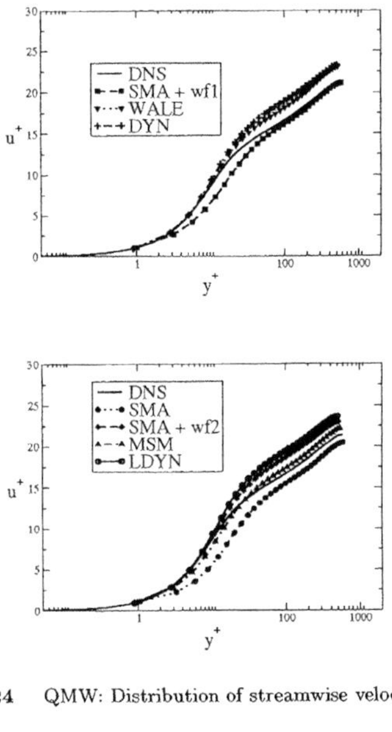

Figure 24 QMW: Distribution of streamwise velocity u^+ vs y^+

Most channel-flow simulations were performed for $Re_b = 10935$, $Re_\tau = 590$, based on channel half width and bulk velocity or wall shear stress, for which DNS data [19]. Some wall-function computations were also undertaken for $Re_b = 21000$, for which only LES data are available for reference. The computational box was: $2\pi h \times 2h \times \pi h$. Statistical properties were evaluated by averaging across the homogeneous $x-z$ plane and over a period corresponding to 24 channel lengths.

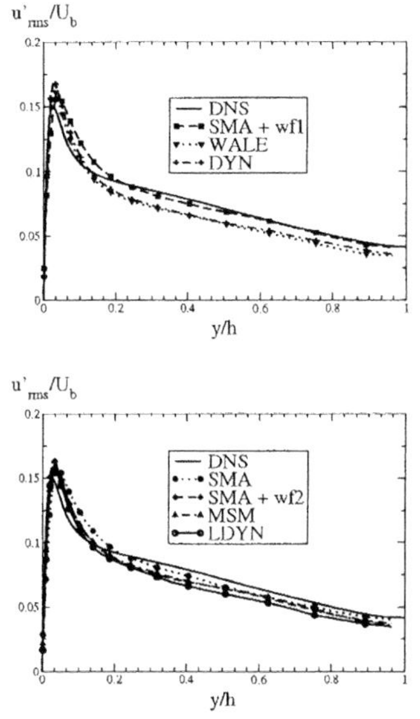

Figure 25 QMW: Distribution of r.m.s. streamwise velocity u'_{rms}/U_b

Simulations were done for the Smagorinsky model, the mixed-scale model (ONERA/LIMSI), the dynamic eddy-viscosity model, the WALE model (CERFACS) and the Localized Dynamic Model [14]. In conjunction with the Smagorinsky model, two different wall-damping functions described in Eq. refwalldampsmag were used.

$$WF1 = (1 - e^{-y^+/5})^2 \quad WF2 = (1 - e^{-y^+/25})^2 \tag{4}$$

Model designations are summarised in Table 7, while Table 9 gives results for wall-friction Reynolds number and maximum mean velocity (for

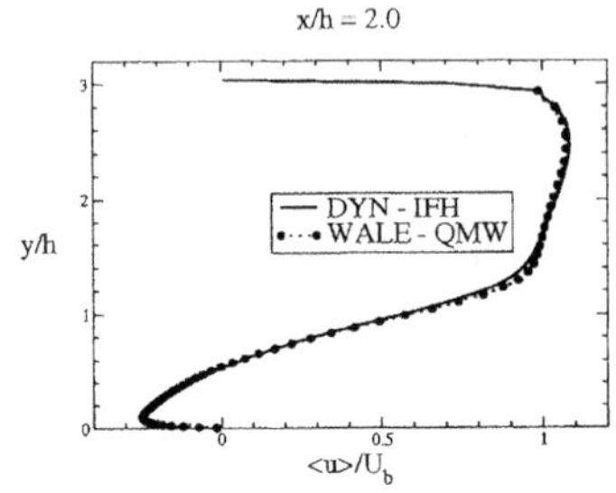

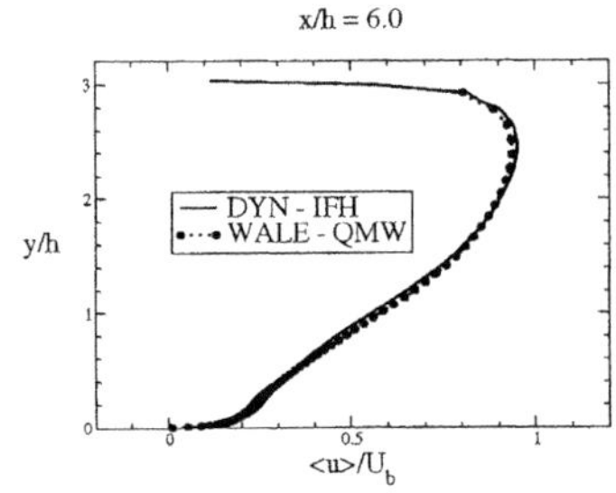

Figure 26 QMW: Distribution of streamwise velocity $\langle u \rangle / U_b$ at $x/h = 2.0$ and $x/h = 6.0$.

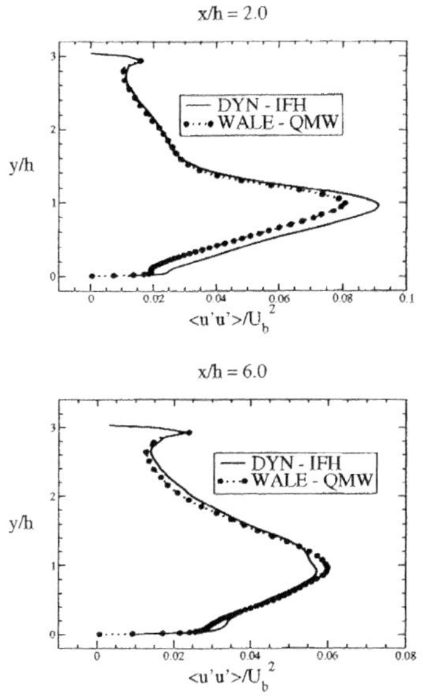

Figure 27 QMW: Distribution of streamwise resolved stress $\langle u'u' \rangle / U_b^2$ at $x/h = 2.0$ and $x/h = 6.0$.

prescribed bulk velocity). These computations have all been undertaken with wall-resolving grids, listed in Table 8, ie without the use of near-wall approximations.

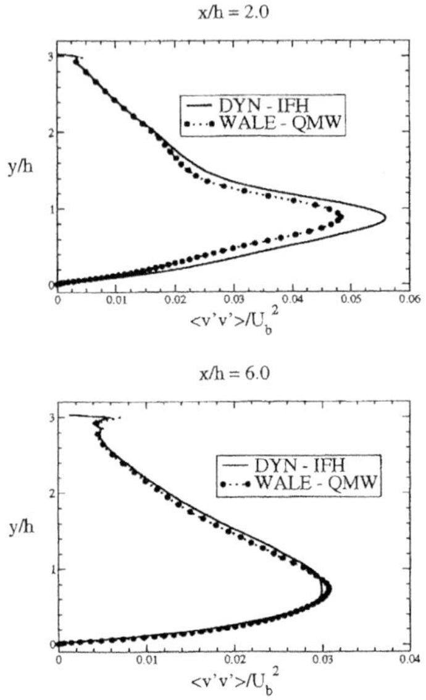

Figure 28 QMW: Distribution of normal resolved stress $\langle v'v' \rangle / U_b^2$ at $x/h = 2.0$ and $x/h = 6.0$.

A major objective of this study was to examine the influence of the SGS viscosity on the agreement with DNS data. Of particular relevance appears to be the asymptotic behaviour of the viscosity close to the wall. As noted already in CERFACS's report, the correct variation is proportional to y^3 and many SGS models do not satisfy this behaviour, either theoretically or in practice when the near-wall region is poorly resolved. The near-wall variation of the SGS viscosity obtained with the various models is shown in Fig. 23. As seen, only three models broadly satisfy the required asymptotic variation, among them the WALE model. This model also gives the lowest SGS viscosity. Curiously, the range of maximum SGS viscosity is found to cover more than an order of magnitude, and this is consistent with CERFACS' finding.

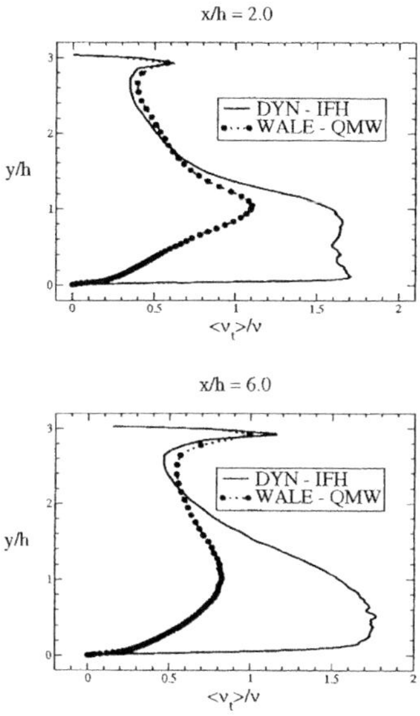

Figure 29 QMW: Distribution of subgrid-scale viscosity $\frac{\langle \nu_t \rangle}{\nu}$ at $x/h = 2.0$ and $x/h = 6.0$.

The errors in wall-shear stress (or associated Re number) and centre-line velocity relative to the bulk velocity are presented in Table 9.

Velocity profiles, in wall coordinates, are given in Fig. 24, while Fig. 25 shows some streamwise intensity profiles. Overall, reasonable agreement with the DNS data is achieved, with some models giving fairly close agreement with the DNS data. The study has brought out, however, the importance of modelling the semi-viscous sublayer. The Smagorinsky model, without any wall damping, produces the wrong behaviour at the wall, but gives results closer to DNS in terms of turbulence energy level. If strong damping is applied to the near-wall viscosity, the wall-law is recovered well, but the skin friction is underestimated, the streamwise normal stress is overpredicted just

Table 7 Subgrid-scale models.

Model Designation	Model Description
Smag	Smagorinsky ($Cs = 0.1$)
Smag + wf1	Smagorinsky ($Cs = 0.1$) + Wall-damping function 1
Smag + wf2	Smagorinsky ($Cs = 0.1$) + Wall-damping function 2
MM	Mixed Model (ONERA/LIMSI)
Germ	Dynamic eddy-viscosity model (Germano), filtering using the trapeze rule
GLT	Dynamic eddy-viscosity model (Germano-Lilly), filtering using the trapeze rule
GLS	Dynamic eddy-viscosity model (Germano-Lilly), filtering using the Simpson rule
GLNT	Dynamic eddy-viscosity model (Germano-Lilly), filtering using the trapeze rule, no planar averaging
NSGS	No subgrid-scale model
Wale	WALE Model (CERFACS) ($Cw = 0.1$)
LDYN	Localised Dynamic eddy-viscosity model (Piomelli & Liu)

outside the viscous sublayer and the turbulence energy in the channel centre is underestimated. Results obtained with the dynamic models show the influence of grid resolution and of test-filtering. The use of 3 different grids in conjunction with the dynamic model shows the necessity of using denser grids in order to obtain results comparable to other subgrid-scale models. All models used in conjunction with coarse grids lead to poor results. Poor resolution manifests itself, among others, by excessive streamwise stress and too low wall-normal stress. This is in line with observations by other partners. A computation without subgrid-scale model has shown that the influence of the numerical error is not negligible. Overall, the best results are achieved with the mixed model without damping, with the Smagorinsky model with WF2, giving high damping or with the WALE model.

Table 8 Description of the grids.

Grid	Number of Cells	Δx^+	Δy^+	Δz^+
1	$72 \times 48 \times 32$	51.5	$10 - 37.1$	58
2	$96 \times 64 \times 64$	38.6	$2 - 42.2$	29
3	$96 \times 64 \times 96$	38.6	$2 - 42.2$	19.3

Wall-function simulations were undertaken with the 4 approximations listed in Table 10 for $Re_b = 23700$, $Re_\tau = 1050$, on a Grid of $50 \times 20 \times 10 = 10^4$ cells with $\Delta x^+ = 165$, $\Delta y^+ = 105$, $\Delta z^+ = 165$. Three of the wall-functions are based on log-law representation of the instantaneous near-wall velocity and the 4th is the Werner-Wengle formulation also used by Karlsruhe and CERFACS. All four approximations have been examined in conjunction with the WALE model, found earlier to return favourable performance with wall-resolving grids. Computations are compared with DNS data [19]. The choice of a higher Reynolds number than in the previous section was motivated by two considerations. The first is the need to maintain a reasonable number of points in the y-direction, whilst having a uniform grid suitable for wall functions ($\Delta y^+ \approx 100$). The second is the wish to make a direct comparison of our results with those obtained by CERFACS in order to complement their extensive study. Result are presented in Table 11, in the same format used previously.

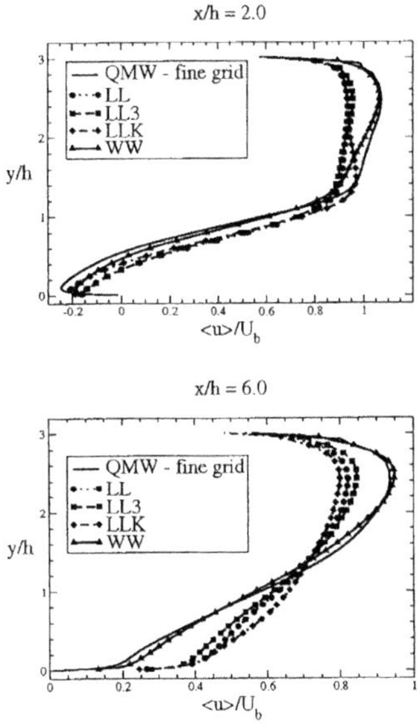

Figure 30 QMW: Distribution of streamwise velocity $\langle u \rangle / U_b$ with the WALE model for different wall-treatments at $x/h = 2.0$ and $x/h = 6.0$.

Overall, agreement is reasonably satisfactory, although, as also shown by Karlsruhe and CERFACS, no wall-function treatment gives a satisfactory description of the structure of the near-wall layer.

Table 9 Predictions of wall shear stress and centreline velocity.

SGS Model	Grid	Re_τ	Error to DNS	u_c/U_b	Error to DNS
DNS	-	584	-	1.1418	-
Smag	2	617	+5.6 %	1.1533	+1.00 %
MM	2	578	-1.0 %	1.1424	+0.05 %
Smag + wf1	2	595	+1.8 %	1.1444	+0.23 %
Smag + wf2	2	538	-7.8 %	1.1319	-0.86 %
GermT	1	463	-20.5 %	1.1165	-2.22 %
GermT	2	520	-10.8 %	1.1216	-1.77 %
GLT	1	467	-19.8 %	1.1231	-1.64 %
GLT	2	523	-10.3 %	1.1239	-1.57 %
GLT	3	551	-5.6 %	1.1253	-1.45 %
GLS	2	529	-9.3 %	1.1298	-1.05 %
GLNT	1	469	-19.5 %	1.1232	-1.63 %
Wale	1	474	-18.6 %	1.1257	-1.41 %
Wale	2	542	-7.1 %	1.144	+0.2 %
LDYN	2	N.A.	N.A.	N.A.	N.A.
NSGS	1	481	-17.5 %	1.1234	-1.61 %

For the hill geometry, a large number of simulations were performed on the coarse and intermediate grids, with the main aim being to investigate aspects of modelling the near-wall region with different near-wall laws. To enable performance to be judged, benchmark data were generated by undertaking a simulation on the finest 'gold' mesh (Grid 3). This almost fully-resolved computation had previously only been performed by Karlsruhe because of its very high cost. A repeat of the simulation, with a different code to that used by Karlsruhe, was motivated by the desire to cross-check the validity of these crucially important results. Also, UMIST/QMW intended to generate its own database for $a - priori$ testing of models and near-wall treatments, and to investigate issues such as realisability and the stress budgets. The UMIST/QMW simulation was performed, in parallel computing mode, on 256 processors of the Manchester University Cray T3D and consumed about 20000 processor hours (about 100 wall-clock hours). As shown in Figs. 26-28, the results were found to agree fairly closely with those obtained by Karlsruhe. The differences observed in respect of stresses are assumed rooted in the different SGS models used by Karlsruhe (dynamic) and UMIST/QMW (WALE). An indication of the differences in SGS viscosities is given in Fig. 29, and these concur with observations made by UMIST/QMW based on channel-flow simulations. The data generated were used, as intended, to examine realisability (via the Lumley invariants map) at different streamwise locations, and of the budgets of the second-moment and the turbulence energy equations. These will be reported in one of two companion papers currently under joint preparation by Karlsruhe and UMIST/QMW and in Temmerman & Leschziner [34]. The data are also used for examining the predictive performance of advanced RANS closures in a project unrelated to LESFOIL.

Table 10 Description of the wall functions.

Denomination	Description
ILL	Instantaneous Log-Law
WW	Werner-Wrengle

The simulations undertaken on the coarse and medium grids are summarised in Table 12. These cover three parametric axes: grid density, SGS modelling and near-wall treatment. The combinations adopted were based on observations derived from channel-flow computations. As pointed out by Karlsruhe, a problem with this flow is that, even for the coarsest grid, the y^+ value along the grid line closest to the wall is mostly below or not much above 11. This does not, therefore, allow an entirely effective or representative examination of wall laws, simply because the Reynolds number of the flow is low. The grids were generated by C. Mellen, Karlsruhe.

Table 13 gives results for the time-averaged separation and reattachment points, relative to the fine-grid simulations. As seen, there are some major differences relative to the correct values and substantial sensitivity to grid density, subgrid-scale modelling and near-wall treatment while. Figures 30-32 give some results reinforcing the sensitivity of the predicted solution to variations in the near-wall treatment.

Table 11 Predictions of wall shear stress and centreline velocity.

SGS Model	Re_τ	Error	u_c/U_b	Error
Expected	1050	-	1.11	-
Wale + LLI	1160	+10.5 %	1.10	0.9 %
Wale + WW	1166	+11.0 %	1.103	0.6 %

In addition to the above studies, some a-priory tests were conducted in which the validity of the wall-laws was examined by reference to time-averaged near-wall velocity profiles computed on the finest grid. Thus, the log-law profile were compared with the profiles predicted by the highly-resolved simulation. Moreover, the wall shear stress predicted by the fine-grid simulation was compared to that arising from wall laws applied to the region lying

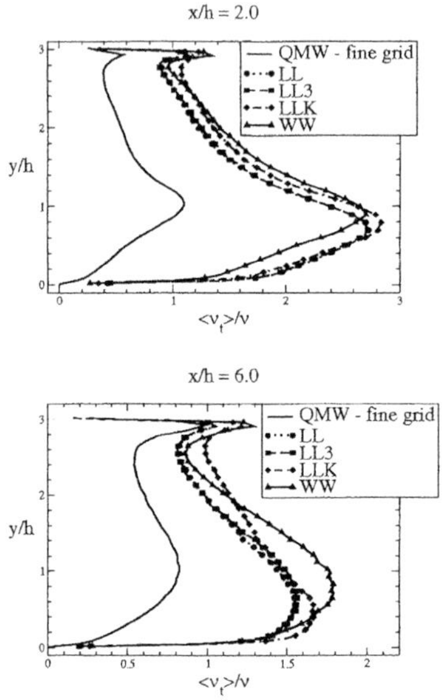

Figure 31 QMW: Distribution of of subgrid-scale viscosity $\frac{\langle\nu_t\rangle}{\nu}$ with the WALE model for different wall-treatments at $x/h = 2.0$ and $x/h = 6.0$.

between the first, second or third grid line closest to the wall and the wall itself. An example is shown in Figure 33 which pertaines to the flow near the lower wall. The results indicate that, in this relatively low-Re flow, the near-wall region does not comply even remotely with the log-law, at any location along the lower wall. This adds weight to a comment made earlier about this flow not being ideally suited to an examination of the efficacy of wall laws.

Table 12 Description of the computations

Ref.	Grid/Computation	SGS Model	Wall Treatment
A	3/Karlsruhe	Dynamic Model	No-slip
1	3/QMW	WALE	No-slip
2	2/QMW	WALE	No-slip
3	2/QMW	WALE	Werner-Wengle
4	2/QMW	WALE	3-Layers Log Law
5	1/QMW	WALE	No-slip
6	1/QMW	WALE	Werner-Wengle
7	1/QMW	WALE	2-Layers Log Law
8	1/QMW	WALE	3-Layers Log Law
9	1/QMW	WALE	Log Law using k
10	1/QMW	SMA + WF2	Werner-Wengle
11	1/QMW	MSM	Werner-Wengle
12	1/QMW	DYN	Werner-Wengle
13	1/QMW	LDYN	Werner-Wengle
B	1/Karlsruhe	Dynamic Model	Werner-Wengle

2.7 Conclusions and Overall assessment

A substantial amount of work has been done on fundamental flow geometries by 4 partners towards identifying optimal subgrid-scale and near-wall modelling practices on the basis of simulations for two flows, one attached and the other separated. These studies formed the foundation for decisions taken in relation to airfoil simulations at a much higher Reynolds number. In total, some 80 simulations were performed for different combinations of grids, SGS models and near-wall approximations. Of these, the generation of bench-mark data for the hill flow by Karlsruhe and UMIST/QMW deserve to be highlighted as being by far the most resource-intensive effort, involving many thousands of supercomputer CPU hours. This activity had not been anticipated at the start of the programme. It was necessitated by the conclusion that the experimental data base, originally expected to be exploited, was unsuitable and did not meet stringent accuracy constraints.

CERFACS has applied an wide range of practices to the channel flow, focusing, in particular, on the near-wall behaviour of subgrid-scale models.

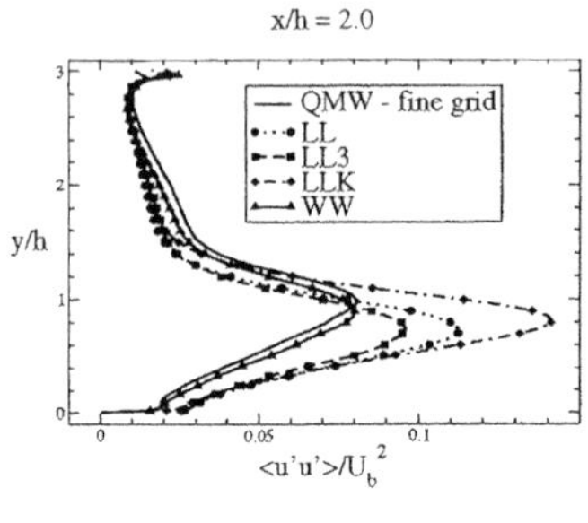

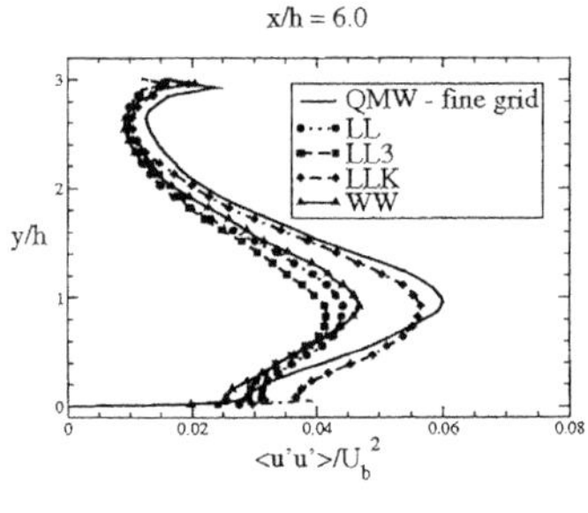

Figure 32 QMW: Distribution of streamwise stress $\langle u'u' \rangle / U_b^2$ with the WALE model for different wall-treatments at $x/h = 2.0$ and $x/h = 6.0$.

From these studies, important lessons have emerged on the relationship between the asymptotic near-wall behaviour of alternative models and their predictive performance. Paradoxically, the crudest model – that due to Smagorinsky – was found to give among the most favourable predictive characteristics – but, importantly, for the wrong reasons. This model is overly diffusive and does not satisfy the requisite near-wall behaviour. Dynamic models and WALE-type model do so, but cannot provide an adequate description of the near-wall region if resolution is poor. Wall-function practices have been shown to give fair results, but significant errors arise if the wall-function layer is thick and resolution is poor.

Table 13 Influence of the grid density on the size of the recirculation bubble

Ref.	$(\frac{x}{h})_{det.}$	$(\frac{x}{h})_{reat.}$	Grid size $(N_X \times N_Y \times N_Z)$
5	1.12	2.17	$112 \times 64 \times 92$
2	0.38	3.45	$176 \times 64 \times 92$
1	0.22	4.72	$196 \times 128 \times 186$

Chalmers has demonstrated the viability of combining a RANS $k - \omega$ near-wall model with the outer LES flow in which the subgrid stresses were approximated with a one-equation SGS k_{sgs} model. Particular attention was placed on the conditions at the matching line. This hybrid LES-RANS methodology was applied to both the channel flow and the hill flow. The mean flow was reasonably well predicted, both for the channel flow and the hill flow, provided the matching line was placed in the inner part of the log-law region. Kinks in the velocity profiles were observed in the region of the matching line, but this problem was much less pronounced than in the hill flow. Based on the current state of knowledge derived from the study, low-Re modelling in the near-wall layer, whilst expected to be fundamentally sounder than wall functions, do not appear to offer outstanding predictive advantages over wall-function approximations. However, more research is needed in relation to model sensitivity and the influence of different matching practices to arrive at a secure conclusion on RANS/LES hybrids.

Karlsruhe's channel computations have confirmed that the Werner-Wengle wall model gives acceptable performance for attached flows, returning the correct wall shear stress. Accuracy declines, however, as wall normal resolution is decreased, though it is difficult in these circumstances to isolate the role of the near-wall model from those of the SGS model and the error arising from the discretistation. This corresponds to recent findings of Cabot and Moin [35]. In terms of turbulence quantities, wall function computations give only moderately acceptable predictions of the near-wall fluctuations. However, this is to be expected it if coarse grid are used. These simply cannot resolve the strong turbulence gradients and the maxima that arise near the wall. In the case of the hill flow, an assessment of the performance of Werner-

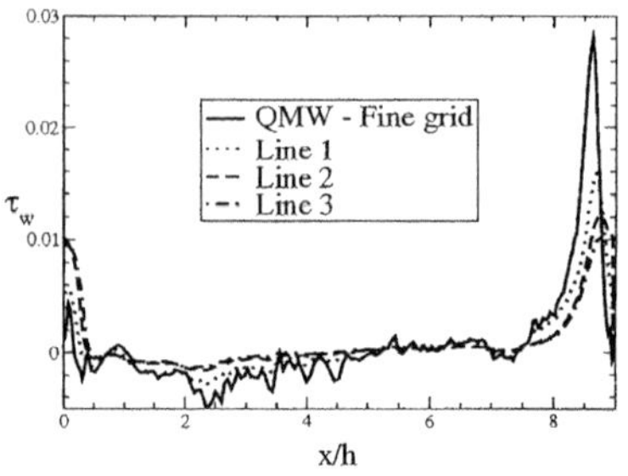

Figure 33 QMW: Distribution of wall shear stress along the lower wall computed from the fine grid computation and via *a-priori* tests using a log-law representation at 3 different locations from the wall

Wengle wall-function approximation is made difficult by the fact that, even with the coarsest grid, the y^+ value along a significant proportion of the near-wall grid plane is below 12. In such circumstances, a no-slip condition replaces the Werner-Wengle wall law. Overall, reasonably good predictions of the separation zone are returned if resolution is adequate. Use of the new backflow wall model failed to give results appreciably superior to that of the Werner-Wengle wall law. This is primarily due to the same reason, the relatively small y^+ encountered in the periodic hill wall function meshes.

Table 14 Influence of the wall treatment on the recirculation zone length

Ref.	Grid	$(\frac{x}{h})_{det.}$	$(\frac{x}{h})_{reat.}$	Wall Treatment
5	1	1.12	2.17	No-slip
6	1	0.46	4.0	Werner-Wengle
7	1	0.54	2.95	2-Layers Log Law
8	1	0.53	2.98	3-Layers Log Law
9	1	0.49	3.38	Log Law using k
3	2	0.32	4.56	Werner-Wengle
4	2	0.34	4.32	3-Layers Log Law

UMIST/QMW wall-resolving simulations identified the WALE model as giving close to the best representation of the DNS solution. They also confirmed the importance of adequate resolution of both the mean-flow velocity and Reynolds stresses. In consonance with Karlsruhe's and CERFACS' experience, UMIST/QMW has also concluded that wall-function approximations give an adequate statement on coarse meshes, but that the turbulence field is progressively poorly resolved as the near-wall layer is increased. For the hill flow, UMIST/QMW has demonstrated the strong sensitivity of the simulation results to subsgrid-scale modelling, grid density and wall-layer approximation. Insufficient streamwise resolution in the vicinity of the separation point is especially damaging to accuracy. With this problem addressed through streamwise refinement, the combination of the WALE model with the Werner-Wengle wall function gives an entirely credible representation of the fine-grid solution, both in terms of mean velocity and turbulence quantities. This is in particularly stark contrast to the adoption of the no-slip condition at the wall.

Table 15 Influence of the SGS Model on the recirculation bubble length

Ref.	Grid	$(\frac{x}{h})_{det.}$	$(\frac{x}{h})_{reat.}$	SGS Model
6	1	0.46	4.0	WALE
10	1	0.5	3.59	SMA + WF2
11	1	0.45	4.18	MSM
12	1	0.46	3.56	DYN
B	1	0.5	3.20	Dynamic Model
13	1	0.47	3.56	LDYN

3 Task 4: Numerical methods

Task Leader: Fluent
D. Cokljat

3.1 Introduction

During the LESFOIL project a variety of numerical methods were tested. Indeed, there were no two partners in the project who had identical numerical procedures. However, despite differences in the underlying technologies, all of the research and industrial groups aimed to improve those features in their respective methods that were relevant to performing LES. This basically meant fulfilling two principle requirements:

- the methods must be as efficient as possible in order to reduce CPU costs for the LES runs so they can become more attractive for everyday industrial usage;
- the methods must also be accurate enough to prevent numerical errors significantly affecting the accuracy of the LES predictions.

In order to achieve these two goals all the partners set out to improve their existing numerical procedures. Given the versatility in numerical methods used by the partners, each partner got engaged in different activities which can be summarized as follows:

- Chalmers initially used their time to evaluate solver performances on the backward-facing step and on the airfoil. They evaluated different pressure correction methods and parallelized their code using PVM [36]. The solver performances on these two test cases showed very satisfactory results.
- The University of Surrey has developed and adapted the contravariant-velocity LES code on a staggered mesh. This they used with a C mesh about the A- airfoil and also for studying the test case of separated flow in a shaped channel. The code has been parallelized using MPI.
- UMIST has developed a new general multi-block LES code specifically for the LESFOIL project. The code has been parallelized using PVM and MPI for the Cray T3D platform. It was applied to the fully-developed channel flow and A-airfoil using up to 256 processors on the Cray T3D. Code evaluations indicated very good efficiency values.
- Fluent used the first half of the project to evaluate the performance of a segregated solver (part of the general-purpose FLUENT V5 solver) on a cold-flow triangle stabilizer. The results presented here indicate the best numerical strategy for LES simulations. The numerical method of this solver has been previously parallelized [37]. Furthermore, Fluent spent some time on analysing impact of different discretisation schemes on the accuracy of the LES results.

- During the LESFOIL project, ONERA developed a local mesh refinement technique which significantly enhanced the capability of LES to predict the flow around an isolated airfoil on the available supercomputer.
- University of Karlsruhe devoted significant effort to reducing the cost of computing the airfoil test case. As a result, the efficiency and capabilities of the LESOCC code have been significantly enhanced. The major modifications are described below and include the development of Fourier solver (technology transfer from University of Surrey) and also development of 2D/3D zonal refinement method.

Following sections describe in detail the activities that all partners have conducted in order to improve their numerical methods.

3.2 Contribution by Chalmers

Authors: S. Dahlström, L. Davidson

The code that has been used is CALC-BFC, a finite-volume Navier-Stokes solver with a collocated grid arrangement [38]. This code was parallelized using multiblock and PVM [39] and recently it has been modified further to use MPI. For the advancement in time, the Crank-Nicolson scheme was used and the momentum equations were discretised in space using central differencing. The central difference scheme was implemented using the deferred correction approach (see below). The CGSTAB-solver was used for the velocities and the SIP-solver was used for the pressure correction equation. For the pressure-velocity coupling the PISO algorithm was used. The PISO algorithm consists of iterative steps, where the first step is exactly the same as the one in the SIMPLE algorithm.

Performance Assessment. The flow over a backward-facing step was used as a test case. The step is $2.5h$, where h is the inlet height. Two different meshes were employed, one coarse ($130\times82\times42$ nodes) and one finer ($130\times82\times82$ nodes). The width of the computational domain is $18h$ and the extension downstream the step is $69h$.

The flow at a Reynolds number of 5000, (based on the inlet velocity and inlet height) was computed using LES with the Smagorinsky subgrid model ($C_S = 0.1$). For advancement in time, the Crank-Nicolson scheme was used, except for the pressure, which here was treated fully implicitly. A 1/7-profile was used as inlet data and a convective boundary condition was used at the outlet. The no-slip condition was used at all walls. The time step was set to 0.26 seconds, which gave a maximum CFL number smaller than 1.2.

Speed-up. A converged $k-\varepsilon$ solution was taken as an initial solution for the LES. The effectiveness of the message passing PVM code was tested using the

k-ε model on the finer mesh. With the domain divided into one, four and eight sub domains, the speed-up (measured in elapsed time to reach convergence) was found to be 4.3 and 8.3 on four and eight processors respectively. Here the momentum equations were discretised using the hybrid difference scheme. In this simple case, the speed-up was approximately linear.

Deferred correction. LES computations using the SIMPLEC algorithm for the pressure-velocity coupling and on the coarse mesh, resulted in a higher rate of convergence when the deferred correction approach was used. For various under-relaxation parameters for the velocities, the number of iterations per time step decreased by approximately 10% (see Figure 34). Eight

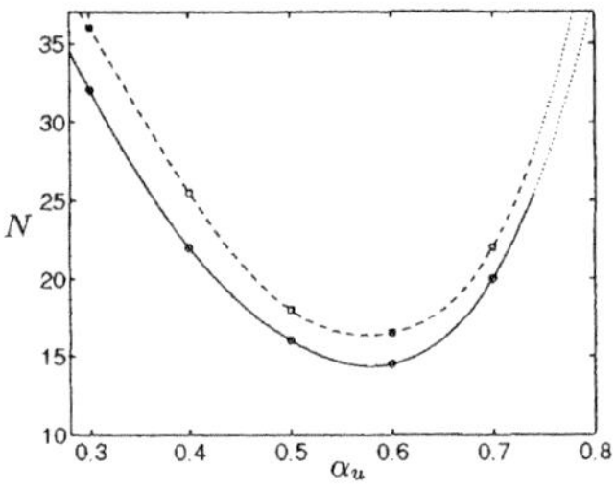

Figure 34 Number of iterations, N, per time step versus the under-relaxation parameter for the velocities α_u ($= \alpha_v = \alpha_w$; $\alpha_{p'} = 1$). Circles: the deferred correction approach; squares: the modification in Ref. [40].

processors were used in these computations.

PISO and SIMPLEC. In the computations with eight processors on the coarse grid, the number of iterations per time step was found to be approximately 5.4 with the PISO algorithm. Although the PISO algorithm is more time consuming per iteration than the SIMPLEC algorithm (in this case approximately 30% more expensive), the CPU-time per time step for the PISO algorithm (with no under-relaxation) was reduced by approximately 50%, compared to the optimized SIMPLEC algorithm ($\alpha_u = 0.6$ in Figure 34). There were no inner iterations of the second corrector step in these computations and further evaluation can be done in optimizing the PISO algorithm. The CGSTAB-solver was used for the velocities and the SIP-solver was used for the pressure corrections.

Spatial Discretisation When central difference approximations are used for the convective fluxes in the momentum equations, the equation systems may not be diagonally dominant, leading to no convergence. One way of handling this problem was presented in [40], and another way is *deferred correction* [41]. Given an equation system

$$a_p u_{i,p}^m = \sum_{nb} a_{nb} u_{i,nb}^m + Q$$

where nb stands for neighboring nodes, the modification in [40] consists of adding $\sum_{nb}(|a_{nb}| - a_{nb})$ to the diagonal coefficients a_p and $u_{i,p}^{m-1} \sum_{nb}(|a_{nb}| - a_{nb})$ to the right hand side Q, where $u_{i,p}^{m-1}$ are the velocities from the previous outer iteration $(m-1)$. Deferred correction implies that the convective flux could be expressed as

$$\dot{m} u_{UP}^m + \dot{m}(u_{CD}^{m-1} - u_{UP}^{m-1})$$

where CD stands for the higher order central differencing scheme and UP for the lower order upwind scheme.

Provided that the equation system converges, both approaches lead to the accuracy of the central differencing scheme.

3.3 Contribution by Fluent

Author: D. Cokljat

The numerical methods employed in this study are part of general purpose code Fluent V5. It is based on a cell-centered, finite volume discretisation and a SIMPLE-type segregated solution procedure. Discretisation schemes are second order in both space (central differencing) and time (second order implicit). The computational domain is discretised into arbitrary unstructured convex polyhedral over which governing equations are integrated. This leads to the set of linear algebraic equations which are consequently numerically solved. Full details about numerical method used here are available in [42].

Performance Assessment. The numerical algorithms available in FLUENT V5 were used to calculate cold flow around a triangle stabilizer. The grid used for this test was two-dimensional with 26000 cells. This test case was chosen because it had been previously used in [43] where it was shown that LES produces superior results (despite being a 2D simulation) compared to a more traditional RANS approach. In order to evaluate the real CPU time required for LES simulations we will analyses efficiency of the full LES algorithm (pressure + momentum).

LES simulations were started from converged $k-\varepsilon$ results. A series of tests were conducted in order to establish efficiency of particular algorithms in the context of LES simulations. A mean velocity of $10m/s$, superimposed with Gaussian random fluctuations of 2% on the mean axial velocity, was specified at the inlet. Wall functions were used at the walls.

The under-relaxation was set to 0.5 for the pressure and to

Table 16 CPU timing per 100 time steps.

CPU (Smagorinsky)	SIMPLE	SIMPLEC	PISO
TEST 1	114	123	160
TEST 2	N/A	84	81

0.3 for the momentum equation. Second-order discretisation schemes were used in both, time and space. The time step was set to 0.0001 and the results were considered as converged at each time step once residuals dropped below the value of 0.001. Main convergence criteria (residuals below 0.001) were chosen in order to ensure accurate solutions during time-advancement so that statistics which will eventually be taken will be reliable. Table 16 gives CPU time (minutes) per 100 time steps for the above conditions (TEST 1). The best results in TEST 1 were obtained when using the SIMPLE algorithm. This can be explained by the fact that both SIMPLEC and PISO are known to produce their best results with higher values of under-relaxation.

TEST 2 followed on from TEST 1 for another 100 time steps but with pressure+momentum under-relaxation increased to 0.9. The increase in under-relaxation did lead to improved speed for both the SIMPLEC and PISO algorithms. It is interesting to note that although PISO required less iterations per time step (typically 8) compared to SIMPLEC, it used almost the same CPU because more work was required per iteration.

In conclusion, it seems that both algorithm, PISO and SIMPLEC, lead to a similar numerical performance for LES simulations using FLUENT V5.

Discretization Scheme. Fluent implemented a central differencing scheme into a general unstructured control-volume based solver. In this approach, a second-order accurate central differencing scheme calculates face value (f) as follows:

$$\Phi_f^{CD} = \frac{1}{2}\,(\,\Phi_0 + \Phi_1\,) + \frac{1}{2}\,(\,\nabla\Phi_{r,0}\cdot \boldsymbol{dr0} + \nabla\Phi_{r,1}\cdot \boldsymbol{dr1}\,)$$

where indices 0 and 1 refer to the cells that share face f, $\nabla\Phi_{r,0}$ and $\nabla\Phi_{r,1}$ are the reconstructed gradients at cells 0 and 1 respectively and finally, $\boldsymbol{dr}$ is the vector directed from the cell centroid towards the face centroid, as shown in Figure 35. It is well known, that CD scheme can produce unbounded solutions and non-physical wiggles which can cause a numerical stability

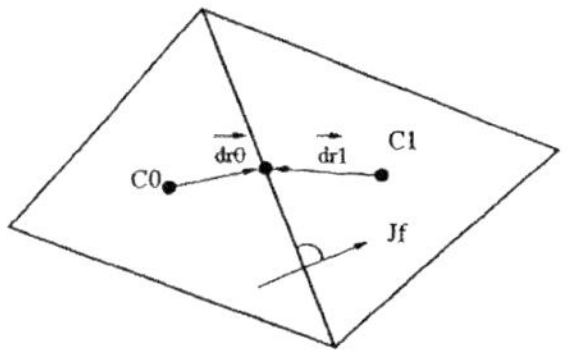

Figure 35 Interior control volume.

problem. Stability problems can be partly avoided if a deferred approached is used for implementation of the CD scheme, as described in the previous section.

Accuracy assessment. The flow around a surface-mounted cubical obstacle placed in a fully developed channel flow was chosen as a test case in order to assess the impact of the discretisation scheme on the accuracy of the LES results. This test case has been one of the standard test cases for a couple of Workshops on Large-Eddy Simulations and turbulence modelling (e.g [44]). Despite quite simple geometry this test has become very popular due to the complex flow structure caused by multiple separation regions and vortices. Detailed experimental data for this test case are available in [45]. Here, we present only basic details about geometry and domain in Figure 36. The cube has a size of $H = 25mm$. The computational domain extends $3H$ upstream and $6H$ downstream from the cube, while laterally it extends $3H$ each side from the mounted cube. The mesh is stretched in the y direction, giving a finer mesh at the top and bottom walls and also the grid around cube is locally refined resulting in a total mesh size of about 900000 cells. Predicted re-attachment lengths at the center-plane for the 3D flow around the cube [46] (normalized by cube height) are given in Table 17. As shown in Table 17, the length of recirculation is strongly influenced by the numerical scheme. The result based on QUICK scheme shows an excessive over-prediction of the re-circulation length by 49% compared to experiment. Many authors believe that level of dissipation of any scheme is even more more important than the formal order of accuracy. This is clearly seen here where the 3rd order accurate QUICK scheme completely fails compared to second-order Central Differencing scheme. This is due to the fact that any upwind-based schemes, including QUICK, generates a large numerical dissipation which acts as an additional sub-grid model. For consistency, we include here LES results published in [46]. It is clear that LES results from [46] are very close to the present ones.

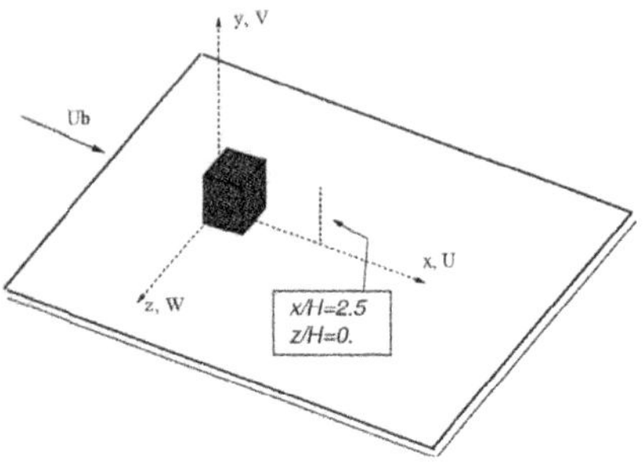

Figure 36 Basic geometry and flow conditions for 3D cube flow.

Table 17 Re-attachment lengths for 3D cube flow using Smagorinsky model.

Method	Re-attachment Length x/H	
	Upstream	Wake
Experiment [45]	1.04	1.61
LES using FLUENT (CD)	1.18	1.78
LES using FLUENT (QUICK)	1.26	2.40
LES by Breuer et al. [46] (CD)	1.228	1.70

Velocity Profiles. Present predictions are compared with the data at six different positions around the cube. Results obtained using CD are consistently in better agreement compared with QUICK results. To illustrate this U and V velocity profiles are plotted at x/H = 2.5, 4.0 in Figure 37.

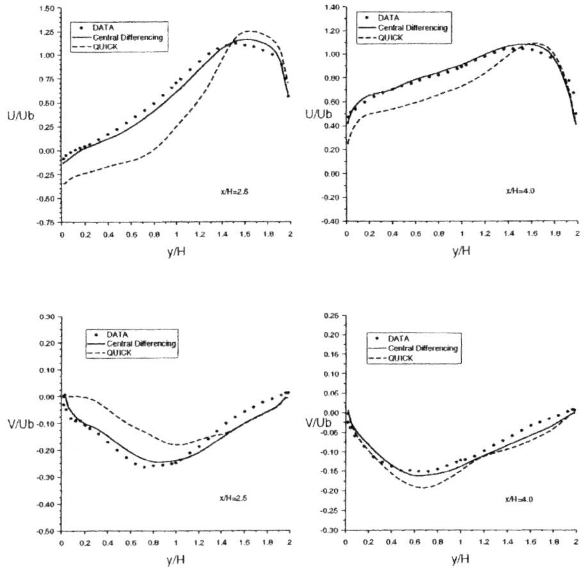

Figure 37 Fluent: Velocity profiles downstream from the cube.

3.4 Contribution by University of Karlsruhe

Authors: C. Mellen, J. Fröhlich, W. Rodi

Fourier solver for the p' equation. The Runge Kutta time stepping scheme used in the LESOCC code requires that at every time step a pressure correction field p' is computed. Previously the p' equation had been solved using the SIP method of Stone. While robust and flexible, this approach is not necessarily optimal when applied to computational domains featuring periodic boundary conditions (as used in the airfoil test case). In this situation an improved p' solver may be obtained by applying a discrete Fourier transform along the periodic spanwise direction. The problem of solving the three-dimensional pressure correction equation

$$\frac{\partial^2 p'}{\partial x^2} + \frac{\partial^2 p'}{\partial y^2} + \frac{\partial^2 p'}{\partial z^2} = -R \tag{5}$$

is then reduced to that of solving N plane two-dimensional problems of the form

$$\frac{\partial^2 \widehat{p'}_k}{\partial x^2} + \frac{\partial^2 \widehat{p'}_k}{\partial y^2} - k_{eff}^2 \, \widehat{p'}_k = -\widehat{R}_k \tag{6}$$

where $k = 0$ to $N-1$ is the wavenumber. The forward and inverse Fast Fourier Transform, respectively, is used to obtain the Fourier coefficients $\widehat{R}_k$ from the residual field R and to then construct the p' field from the $\widehat{p'}_k$ Fourier coefficients. These routines were kindly supplied by the University of Surrey. The transformed problem is the same as the one in physical space if the effective wavenumber k_{eff} of the differencing scheme is applied.

For wavenumber $k \geq 1$, the left-hand side in Eq. 6 is diagonally dominant. Experience has shown that these non-zero wavenumber problems could be very readily solved using a parallelized two-dimensional SIP solver. Solution of the $k = 0$ wavenumber problem proved to be more problematic. Difficulties in maintaining compatibility of the $\widehat{p'}_o$ field across adjacent blocks in the domain decomposition markedly slowed down the rate of global convergence when the 2D SIP solver was applied to this problem. Significantly improved efficiency was obtained when a BiCGSTAB solver was employed in conjunction with Jacobi pre-conditioning for this wavenumber. This combination allows the $\widehat{p'}_o$ field to be treated in a contiguous fashion and automatically ensures compatibility across the entire solution domain.

Comparisons have shown that when applied to a periodic channel geometry the Fourier solver was approximately 3 times more efficient than the standard 3D SIP solver. When applied to the airfoil test case the superior inter-domain coupling of the Fourier method (enabled by use of the BiCGSTAB solver on the $k = 0$ wavenumber problem) yielded much lower residuals. This was important to ensure that the pressure field around the airfoil remained free of numerical artifacts.

Implications of 2D/3D Zonal refinement method on Fourier solver. The approach used by Karlsruhe is the same as originally proposed by

ONERA (see below). However, since the Karlsruhe coupled this technique with Fourier solver in 3rd dimension it is important to stress here that modifications must also be made to the operation of the Fourier solver. In the 2D regions only the coefficients for the $k = 0$ wavenumber are required. The absence of spanwise fluctuations removes the need for the higher wavenumber. In the 3D regions the $k = 1...N - 1$ wavenumber problems still need to be solved however and require boundary values to be provided along the 2D/3D interfaces. Assuming an absence of spanwise variation in the 2D regions, these boundary values are obtained by simply imposing $\widehat{p'}_{k,3D}(x,y) = 0$ (for all $k \geq 1$) in the 3D dummy cells responsible for the 2D/3D information exchange.

3.5 Contribution by ONERA

Authors: I. Mary, P. Sagaut

During the LESFOIL project, a local mesh refinement technique has been proposed by ONERA to enhance the capability of LES to predict the flow around an isolated airfoil on the available supercomputer. Indeed as LES of wall bounded flow requires very small cells near the wall, it becomes essential to optimize the cell distribution in order to limit the total number of grid points. A 2D/3D coupling strategy has been developed in a structured solver, FLU3M, to reduce the total number of grid points. Indeed in such simulations, the flow is laminar and two-dimensional in a large part of the computational domain. Thus three-dimensional computation can be limited to a zone close to the airfoil where the flow is turbulent, whereas two-dimensional simulations are used for the far field (see Fig.38). This strategy, which does not alter the accuracy of the simulation, allows the use of a mesh with LES resolution, while keeping the total number of grid points at a reasonable level. As boundaries conditions are treated through ghost cells, the 2D/3D coupling is realized by the following definitions of ghost cell data:

$$Q_c^{2D}(x,y) = \frac{1}{Lz}\int_0^{Lz} Q_c^{3D}(x,y,z)dz \tag{7}$$

$$Q_c^{3D}(x,y,z) = Q_c^{2D}(x,y) \tag{8}$$

where Lz denotes the spanwise extent of the computational domain. If the boundary between 2D and 3D blocks is located too close to turbulent zones, the flow variables in the 3D block are subject to an implicit spanwise filtering due to Eqs.(7-8). Therefore the extent of the 3D zone in the direction normal to the wall must be chosen carefully to realize a good compromise between efficiency and accuracy. For flow over airfoil, the turbulent boundary layer thickness, which grows from the leading edge to the trailing edge, is often less than 10 percent of the chord. Thus the 2D/3D technique can easily reduce

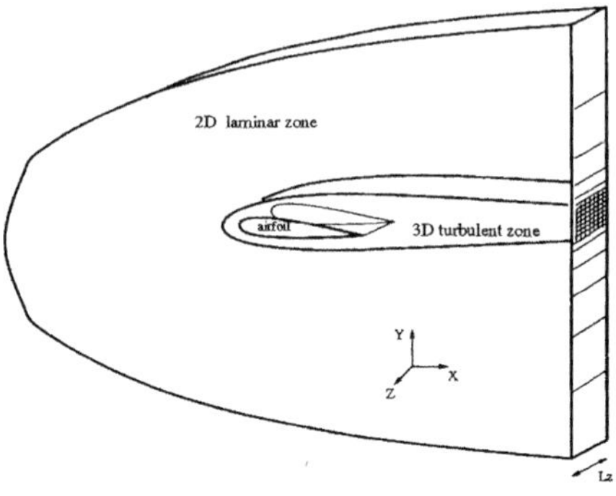

Figure 38 2D/3D coupling strategy.

the total number of grid points by a factor 2, especially if several 3D blocks with different extents in the direction normal to the wall are used to fit the turbulent boundary layer topology.

3.6 Contribution by University of Surrey

Authors: P.R. Voke, L. Jervase, Y. Liu

Surrey have developed and adapted the contravariant-velocity LES code on a staggered mesh for use with the C mesh about the A-airfoil and for studying the test case of separated flow in a shaped channel. The code is fully operational and was parallelized using MPI. Both the basic Smagorinsky model and the dynamic Smagorinsky model were implemented in the fully covariant staggered-mesh code.

In addition to the above, Surrey have developed and implemented a Schur-complement method of pressure solution with excellent parallelization properties, following advice from Prof. Kaltenbach. This was found to solve the low-wavenumber planes of the discrete Fourier transformed Poisson equation for pressure much faster (on a parallel machine) than any iterative scheme we have available, in Cartesian coordinates.

3.7 Contribution by UMIST

Author: L. Temmerman, M. Leschziner

UMIST have developed, verified and applied a new general multi-block LES code specifically for its work in LESFOIL. The code is based on the fully-conservative finite-volume approach in general non-orthogonal co-ordinates with cell-centred collocated Cartesian-velocity components. Time-marching is effected by a fractional-step method. The time-derivative is approximated by a 3-level retarded second-order scheme. Convection and diffusion fluxes are evaluated explicitly using the Adams-Bashforth method. Advancement in time is performed with the Cartesian velocity components, and the Rhie & Chow interpolation method is used to evaluate the mass fluxes at cell faces. The convective terms are computed with a second-order centred scheme to keep artificial dissipation to a minimum.

The pressure field is obtained from the solution of the related Poisson equation. This equation is solved with a partial diagonalisation technique and a multi-grid algorithm.

The code has been parallelized using MPI and PVM for the Cray T3D machine architecture. More recently, the code was successfully transferred to a new machine, the Cray T3E, and its performance on that machine is now similar to that on the T3D which is no longer operational.

Solution of momentum equations. The cell-centred Cartesian velocity components are advanced in time through a second-order, time-retarded 2-step scheme. The first step is explicit, in which a provisional velocity is evaluated by using only the convection and diffusion fluxes. From these provisional cell-centred velocities, contra-variant cell-face velocities are evaluated. The intermediate contra-variant velocity field is then projected onto the space of divergence-free velocity vector field. This is done by correcting the cell-face velocities from the provisional to the final forward-time level using the gradient of the pressure field the value of which is such that a divergence-free mass-flux field arises. Finally, the pressure field satisfying the above constraint is used to update the Cartesian velocity components.

Time-step control. The time-step is variable, which allows the code to operate at optimum speed. In order to retain second-order accuracy, the time derivative and the Adams-Bashforth extrapolation use two time steps. The control of the time step is done by reference to the Courant number and the diffusion number. The maximum values of the two are monitored in order to keep both below prescribed values.

The pressure equation. The pressure equation is derived by combining the relationship between face-velocity corrections and pressure gradient and the mass-conservation constraint. This yields a Poisson equation linking the pressure at forward-time level to the divergence of the provisional, contra-variant cell-face velocities arising from the first, explicit step of the momentum solution.

Domain decomposition and parallelization. Each block is ported to an associated processor of the parallel computer. Each block possesses halo data associated with neighbouring blocks. Whenever necessary, data are exchanged to update the halo region. This is effected by MPI or PVM communication routines. The procedure, originally developed for the T3D machine, is now equally effective on the T3E machine to which the code has recently been ported. This transfer was necessitated by a termination of the T3D service at the Univ. of Edinburgh in May 1999. Access to the new Univ. of Manchester T3E was only enabled in July 1999.

Partial diagonalisation. This is used for accelerating the solution of the pressure Poisson equation. It aims to reduce the number of blocks which are communicating in the multi-grid algorithm. Because of the incompressibility of the fluid, or equivalently, because of the elliptic nature of the Poisson equation, every block has an influence on all others. Thus, the fewer blocks are present the faster the convergence becomes.

For a global domain composed of $N1 \times N2 \times N3$ cells, with the $N3$ direction un-coupled from the others, partial diagonalisation can be employed. For a given grid and given boundary conditions, one may obtain the Eigenvalues and Eigenvectors of the Laplacian operator in the $N3$ direction. Thus, by using the base formed by the Eigenvectors, the linear operator becomes a diagonal matrix (in the $N3$ direction) composed of the Eigenvalues. The 3D problem is then transformed into $N3$ un-coupled 2D problems (one for each Eigenvalue).

Multigrid algorithm. The multigrid algorithm implemented in the solver is a standard V-cycle using a line- relaxation solver and bi-linear interpolation for prolongation. Restriction is effected by agglomeration. Because of the partial diagonalisation method the algorithm needs only to be two-dimensional.

Performance Assessment. The algorithm described above has been verified by reference to several test flows using skewed grids in cavities. It was then applied to LES of fully-developed channel flow and flow around the Aerospatiale A-airfoil using up to 256 blocks distributed on up to 256 processors on the Cray T3D. Efficiency and speedup tests have been performed on a sequence of 32, 64, 128 and 256 processors. Typical efficiency values on 256 relative to 32 processors were found to be of order 85%. Alternatively, the speedup on 256 relative to 32 processors was of the order 7.2 relative to the ideal value of 8. The algorithm now runs satisfactorily on a recently commissioned Cray/SGI T3E computer.

3.8 Achievements and recommendations

Original plan for the numerical methods task was mainly centered around developing and evaluating fast Poisson solvers for LES calculations. However, the original task plan has been slightly modified compared to above. Mainly, some partners (Fluent and Chalmers) modified original plan of development and evaluation of Poisson solvers in favour of pressure-correction method based solvers like PISO and SIMPLEC. This move was justified by the fact that in order to have efficient parallel solver for the Poisson equation it is necessary to use Fourier transformation in the third coordinate direction. This method, despite being very efficient, suffers from limitations which make it very difficult to apply to 3D geometries where there is change in shape in third direction. Consequently this method is less attractive to industrial applications. The partners involved in using pressure-correction methods, Chalmers and Fluent, assessed these methods on series of simple geometries and came up with optimal strategies for using these methods.

The combination of academic methods (Poisson solver with Fourier transformation) and industrially oriented methods, like SIMPLEC and PISO, was found to be beneficial for the project since more methods were tested resulting in more informations being available for the potential LES-user community.

All partners have successfully finished development of their codes and all the codes are in a ready-to-use stage. All codes have been parallelized and each partner has ported them on available platforms whereby MPI has been predominantly used as message passing harness. Evaluation of parallel performances indicated very satisfactory parallel behaviour where in many cases scalability was close to the linear one.

In conclusion, the main aim of the Numerical Methods task, to develop and assess solvers for LES simulations of the flow around an airfoil, has been fully achieved. The main outcome of this task can be summarized as follows:

- Transfer of technology for Fourier transformation in spanwise direction from Surrey to Karlsruhe,
- Comparison of QUICK and Central Differencing (CD) scheme resulted in clear recommendation to use CD for separated flows,
- Evaluation of pressure-correction methods indicated suitability of both, PISO and SIMPLEC, methods for LES runs,
- Development of 2D/3D zonal refinement method by ONERA and transfer of this technique to Karlsruhe.

III. THE AIRFOIL INVESTIGATIONS

III. THE AIRFOIL INVESTIGATIONS

4 Task 5: Airfoil Computations

Task Leader: University of Karlsruhe
C.P. Mellen, J. Fröhlich, W. Rodi

4.1 Introduction

Task 5 is directed towards computations of the flow around the Aerospatiale A-profile at $\alpha = 13.3^o$ and $Re_c = 2.1 \times 10^6$. The objectives in this task were twofold. One goal was to reproduce at best the statistical data obtained in the experiment. The second objective was to investigate the different methodological aspects of LES in the context of an airfoil simulation, such as subgrid-scale modelling, transition models, near-wall treatment and numerical discretization.

All partners have been confronted with the extreme difficulties that the chosen configuration poses for LES computations. The primary source of this difficulty is the large Reynolds number of the flow around the airfoil. At this Reynolds number challenges arise when applying the non-dissipative differencing schemes normally used in LES and when attempting to achieve a mesh resolution capable of meeting the requirements of LES but keeping the computation cost affordable. As might be expected, partners have formulated a variety of novel strategies to cope with these challenges. These outcomes are a reflection of the considerable effort which the partners have devoted to this task.

The Principal Airfoil Geometry. There are very few data sets for incipiently separated airfoil flows at high Reynolds number, which are sufficiently detailed to form a reasonably reliable background against which to compare computational simulations. Such flows present serious experimental challenges, associated mainly with maintaining two-dimensionality and the extreme sensitivity of the flow to even slight disturbances arising from tunnel-interference effects. The Aerospatiale-A airfoil is one geometry which had been judged in a number of earlier RANS studies to qualify as a reliable test case. The geometry is well defined, the experimental data include detailed mean-flow as turbulence quantities across both the airfoil boundary layer and the wake, and experiments were conducted in two windtunnels. While the experiments are inevitably affected by three-dimensional contamination, this is judged to be minor.

The geometry and the coordinate systems used for recording profiles of mean- and turbulence quantities are shown in Fig. 39. Measurements were performed in two different windtunnels, designated "Fl" and "F2", the respective dimensions being 3.5×1.52 and 1.8×1.4 m^2 (height x width). Fl

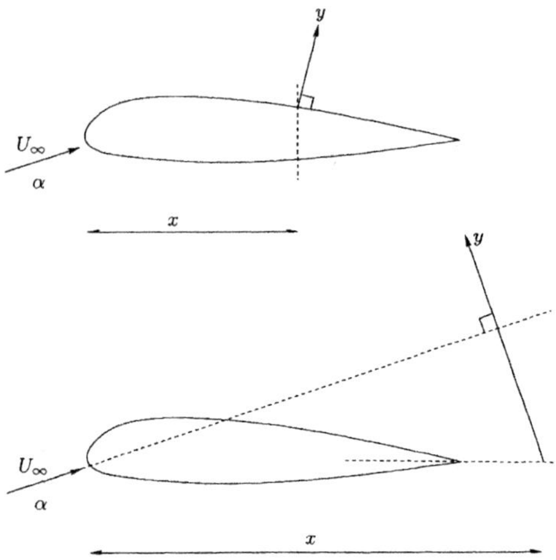

Figure 39 A-Airfoil geometry and coordinate systems for profiles on suction side and wake

was used for global measurements only (drag, lift, pressure and skin friction), while in F2, detailed boundary-layer and wake measurements were performed at incidence 7.2^o and 13.3^o using LDA. In addition, the lift and drag were also measured for other incidence angles. The airfoils used in both windtunnels were geometrically identical (except for a slight difference in the spanwise length), and the Reynolds number, based on the chord, was $2.1 \cdot 10^6$ throughout. The Fl tunnel involved minimal blockage, but slight corrections (less than 1%) to the lift and incidence values were nevertheless applied. This windtunnel also featured an aspiration-based boundary-layer-control system, permitting a high level of two-dimensionality to be maintained for high incidence angles. The smaller F2 tunnel involved somewhat higher blockage effects, and appropriate corrections of the order 2-3% were applied to the measured lift and incidence values. No lateral boundary layer control was available in this windtunnel, and this led to departures from two-dimensionality at incidence angles exceeding 14°. Two-dimensionality in the F2 tunnel was checked by surface oil film visualisation and also by detailed 3D LDA measurements. These measurements suggest that contamination by three-dimensional features were still mild at 13.3^o, especially close to the airfoil's midspan across which the velocity and turbulence measurements were made. The experimental data as well as the geometry can be found on CD-ROM in [2].

In comparisons to follow, data arising from both windtunnels for drag and lift are included. Differences between lift data become pronounced beyond 13°, as the F2 flow begins to be globally affected by 3D features, although, as pointed out earlier, detailed velocity measurements show that the flow is virtually 2D at 13.3^o, which is the most important test condition for present purposes. The experimental investigators also state that the reproducibility of the F2 data is much better than that of the Fl measurements, that the F2 pressure data have been more carefully obtained and that the flow in the Fl tunnel appears to feature transient separation at high incidence above 16°.

Differences in the drag coefficient data are also partly connected to the above issues, but there is the additional fact that drag was determined quite differently in the Fl and F2 flows. In the former, skin-friction strain gauges were used, while in the latter, wake velocity measurements (at 1.5 × chord) were used to compute the drag.

The overall message is that the F1 data should be used to assess the evolution of the computed lift and drag coefficients at high incidence, while the F2 data (at midspan) at 13.3^o provide a good basis for judging the predicted velocity and turbulence fields when the airfoil operates close to stall. This message should be born in mind when considering the simulations to follow.

Common Mesh. The objective of comparing the impact of different submodels for SGS stresses, transition, near-wall flow and of the numerical discretization is most easily attained when different computations are made on the same grid. Comparisons can be undertaken on different meshes but the conclusions reached are mostly not as clear cut. Hence, right from the outset of the LESFOIL project, a task was devoted to grid generation (see Section 1) The grids generated in this task were to be commonly used by all the partners. At first, two structured meshes were generated by ONERA, one supposed to yield wall resolution, the other conceived for use of wall functions. Dassault produced similar unstructured meshes. In a first phase all airfoil computations in the project were undertaken on the common structured meshes but revealed certain deficiencies of these meshes.

It should be noted that with LES the employed grid is part of the turbulence modelling itself due to the implicit filtering it introduces and hence has even more influence on the result than in other CFD approaches. The partners experienced that the generation of a computational grid for LES is difficult to separate from the computation of the flow itself. Rather, an iterative process involving grid generation and simulation is often applied. This is why after first unsuccessful tests the various partners in a second phase generated their own grids and used them in their subsequent computations. Due to the high impact of the mesh with LES a precise assessment and comparison of different methods is rendered more difficult when different meshes are used. It was therefore suggested that all partners perform at least some computations on a common mesh. After the second phase of airfoil computations Karlsruhe presented a thorough comparative evaluation of the different grids employed by the partners. The purpose was to take as much benefit as possible from the computations to date and to identify advantages and shortcomings of the different grids and their impact on the results of the LES. In light of this evaluation and the discussions between the partners Karlsruhe constructed, with technical assistance from Chalmers, two new common meshes which were delivered to the partners.

The meshes generated were structured 'C'-type meshes. The common mesh 'A' contains $720 \times 64 \times 32$ cells in the streamwise, normal and spanwise directions respectively. A coarser 'B' mesh was generated from this mesh by removing every second streamwise point, yielding $360 \times 64 \times 32$ cells. Both meshes have a spanwise extent of $0.12c$ and the near-wall point is located at $y^+ \sim 10$. Of these two, the partners selected the mesh which they felt best balanced their computational resources against the extreme resolution requirements mentioned earlier. Computations on these common grids however led partners to conclude that the resolution they offered was still inadequate for a successful LES. A few partners, Karlsruhe, ONERA, and Chalmers then generated grids which tried to everywhere satisfy resolution criteria typically required for wall-resolving LES within the limits of their computational resources. Resulting from the increased number of points these grids are computationally very expensive. By necessity the spanwise extent therefore had to be reduced with respect to the common meshes for cost reasons. However it was using these grids that the first largely satisfactory LES computations of the airfoil flow were performed. The DES results by FLUENT on the other hand have been obtained on the relatively coarse common 'B' mesh.

4.2 Contribution by Chalmers

Authors: S. Dahlström, L. Davidson

During the LESFOIL project Chalmers' work on the airfoil case has mainly been focused on two issues:

- How should the transition be prescribed?
- If the near-wall region is modelled, what resolution is required in the turbulent boundary layer?

Table 18 Chalmers: Meshes used in the computations.

Computation	Mesh	Spanwise extent, L_z
Smag1	UMIST 321 × 65 × 33	$0.03c$
Smag2	Chalmers 720 × 65 × 33	$0.08c$
Smag3	Chalmers 720 × 65 × 33	$0.08c$
Smag4	Common 721 × 65 × 33	$0.12c$
Yoshi1	Common 721 × 65 × 33	$0.12c$

Numerical Method. The code used (CALC-BFC) is an incompressible finite volume Navier-Stokes solver [38]. The solver is based on structured grids and the use of curvi-linear boundary-fitted coordinates. The grid arrangement is collocated and the Rhie and Chow momentum interpolation method [47] is used. The code is parallelized for 3D flows [36] using block decomposition and the message passing systems PVM and MPI. The PISO algorithm [48] is used for the pressure-velocity coupling, with two additional corrector steps beside the first SIMPLE step. The filtered momentum equations are discretized in time using the Crank-Nicolson scheme and 2nd order difference schemes (the central difference scheme (CDS) and the van Leer scheme) in space. Two SGS models have been used in the computations: the Smagorinsky model and the one-equation SGS model by Yoshizawa [49].

Boundary Conditions. The inlet is specified all over the curved area of mesh facing the inlet flow. The velocities are set to $\overline{u} = U_\infty \cos\alpha$ and $\overline{v} = U_\infty \sin\alpha$, where U_∞ is the freestream velocity. At the outlet, a convective boundary condition is applied: $\frac{\partial \phi}{\partial t} + U_\infty \frac{\partial \phi}{\partial x} = 0$ where ϕ represents any of the velocity components. A Neumann boundary condition is used for the pressure at all boundaries ($\frac{\partial \overline{p}}{\partial n} = 0$). Periodic boundary conditions are used in the spanwise direction for all variables. At the wall, either the no-slip condition or wall functions are used. The wall functions are based on the instantaneous log law in the log region ($y^+ \geq 30$) and a matching between the linear law and the log law in the buffer region ($5 \leq y^+ \leq 30$).

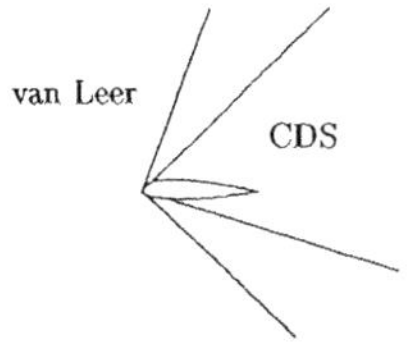

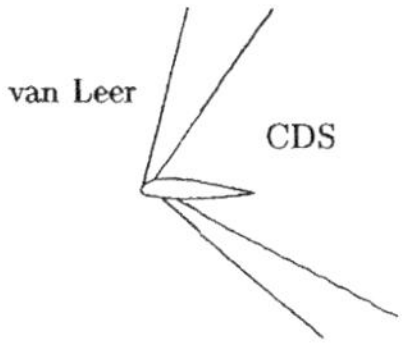

Figure 40 Chalmers: Blending functions used for the scheme in computation Smag1 (left figure) and for Smag2-Smag4 and Yoshi1 (right figure).

Convergence Criteria. For the velocities, the convergence criterion is that the L^1-norm of the residuals of the discretized momentum equations scaled with $\rho U_\infty A_{inlet} U_\infty$ should be less than the desired convergence level, η, where A_{inlet} is the projected area of the inlet. Here, the criterion is checked immediately after the solver (but without under-relaxation), i.e. the residuals are calculated with the 'old' coefficients, and before the correction in the PISO algorithm. For the continuity equation, the criterion is that the L^1-norm of the discretized finite-volume continuity error scaled with the inlet mass flow should be less than η.

Table 19 Chalmers: SGS models used and its treatment in the transition region on the suction side of the airfoil.

Computation	SGS model
Smag1	Smagorinsky
Smag2	no model ($x/c < 0.12$)+Smagorinsky ($x/c > 0.23$)
Smag3	no model ($x/c < 0.12$)+Smagorinsky ($x/c > 0.23$)
Smag4	no model ($x/c < 0.02$)+Smagorinsky ($x/c > 0.12$)
Yoshi1	no model ($x/c < 0.12$)+Yoshizawa ($x/c > 0.23$)

Computations Five computations are presented in this report (Smag1-Smag4 and Yoshi1). Table 18 shows the meshes used in the different computations. The computations have been performed on an IBM SP computer at the Center for Parallel Computing at KTH. The computational domain is decomposed into 8 or 32 subdomains. The elapsed time per time step on 32 processors is $7s$ on the computations on the Common mesh, with the convergence criteria fulfilled after two global iterations ($\eta \approx 1 \cdot 10^{-3}$). The

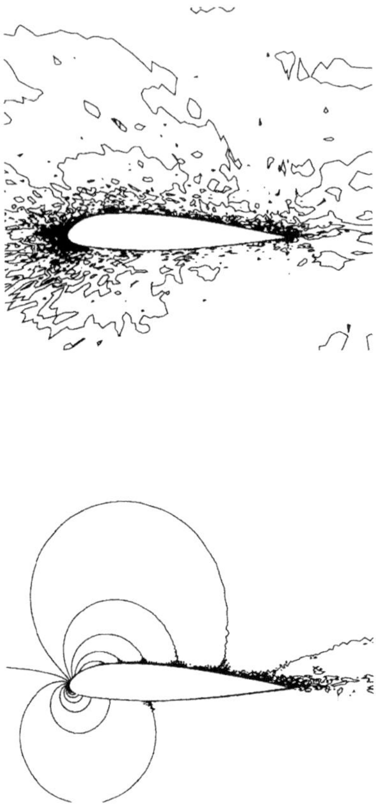

Figure 41 Chalmers: Instantaneous contour plots of the resolved $\bar{u}$-velocity. CDS is used in the entire domain in the top figure and the van Leer scheme is used in part of the domain in the bottom figure.

simulation of one time unit (c/U_∞) requires about 180 CPU hours and with MPI and 32 processors the solution is advanced 4 time units per day (24 hours).

The averaging time is at least 5 time units. The initial conditions are previous simulations or a 2D solution computed with a $k-\varepsilon$ model. The time step is $3 \cdot 10^{-4} c/U_\infty$, giving a maximum CFL number of approximately 1 (0.6 on the UMIST mesh).

Transition. To remove unphysical oscillations in front of the airfoil and upstream of the transition region, a bounded second-order upwind discretization scheme (the van Leer scheme) is used in this region. This scheme is blended with the CDS upstream of the transition point given in the experiments (at 12% of the chord on the suction side). In this way the mixed scheme prevents any numerical transition from taking place where the van Leer scheme is used. In all the computations presented in this report, the blending is between 2 and 12% of the cord on the suction side, except for computation Smag1, where it is between 1 and 9% (see Fig. 40). Figure 41 shows instantaneous contour levels of the $\bar{u}$-velocity when the convective terms are discretised using CDS (top figure) compared to when the second-order upwinding is used in part of the domain (bottom figure). It can be seen that the unphysical oscillations are also damped in the region where the CDS is fully active.

Not only the discretization of the convective terms has an effect on the transition but also the treatment of the SGS models. For example the eddy viscosity from the Smagorinsky model does not disappear in laminar flows (unless the strain-rate tensor is zero). This results in additional dissipation in the laminar region (besides the one from the upwind scheme). In the computations the eddy viscosity is set to zero in the laminar region and gradually mixed with the eddy viscosity from the SGS model. The mixing regions of the SGS models in the computations are summarized in Table 19. For more details, see [50]. Similar approaches are applied around the transition point on the pressure side of the airfoil. In the one-equation SGS model the SGS kinetic energy, k_{sgs}, is set to zero upstream of the transition point and there is no damping or blending in the k_{sgs}-equation.

The Smagorinsky constant, C_S, is equal to 0.1 and the filter width, Δ, is computed using the definition $\Delta^3 = \Delta x \Delta y \Delta z = \delta V$. For the SGS model by Yoshizawa [49], the filter width is taken as the minimum side length of the cell, i.e. $\Delta = \min\{\Delta x, \Delta y, \Delta z\}$. For more details, see [51].

Figure 42 shows the resolved stresses in the streamwise direction for the computations Smag3 and Smag4. The figure illustrates the effect of the SGS models on the transition. A peak in the resolved stresses at the correct location is obtained if the Smagorinsky model is blended downstream of the transition 'point' and there is a delayed transition if the blending is too far upstream. Looking at the skin friction coefficient (Fig. 43) it is obvious that this is the case for the computations Smag1 and Smag4. They exhibit a de-

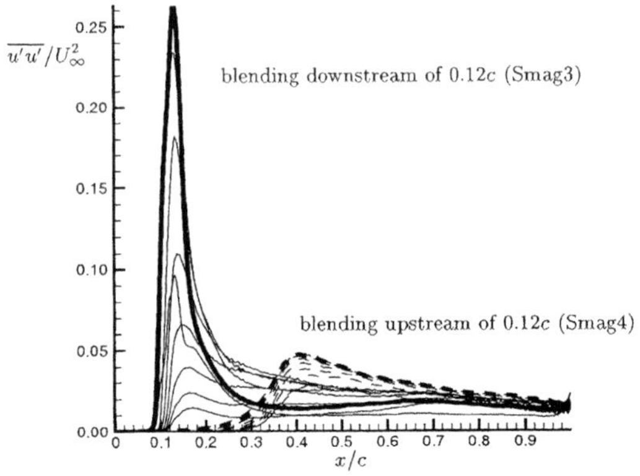

Figure 42 Chalmers: Spanwise and time averaged resolved velocity fluctuations at near-wall nodes along the suction side of the airfoil.

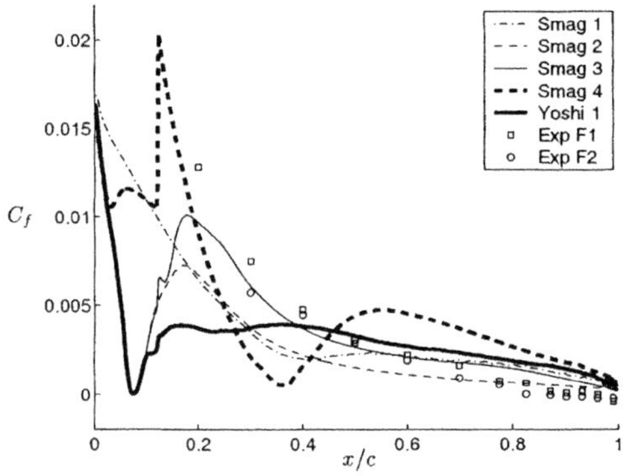

Figure 43 Chalmers: Skin friction coefficient.

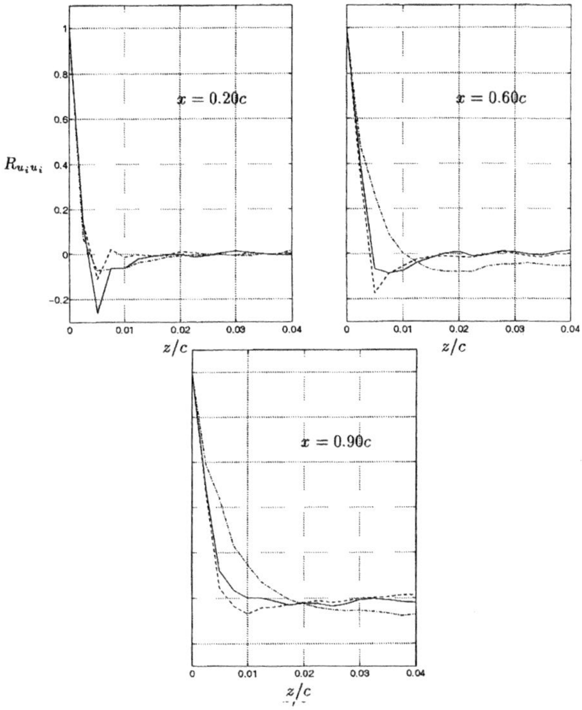

Figure 44 Chalmers: Spanwise two-point correlations $R_{u_iu_i}$ of the resolved velocities at approximately 12% of the boundary layer. Case Smag3. Solid: R_{uu}; dashed R_{vv}; dash-dotted: R_{ww}.

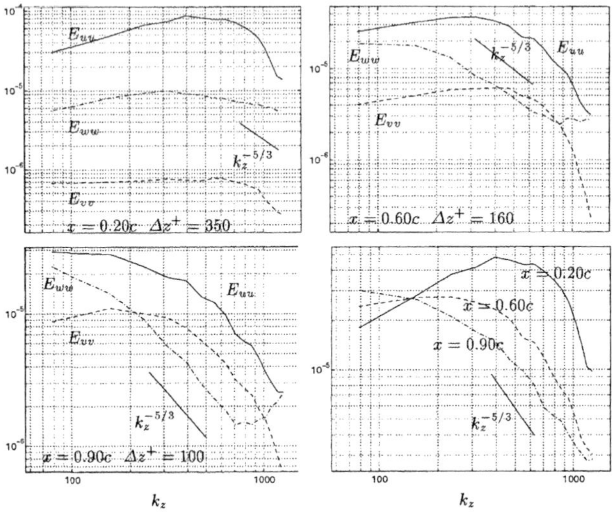

Figure 45 Chalmers: Spanwise one dimensional energy spectra of the resolved velocities at $y/\delta \approx 0.12$ (from Smag3). Lower right figure shows $(E_{uu}+E_{vv}+E_{ww})/2$

layed transition (note that the peak at 12% of the chord for Smag4 is not a sign of transition, it is the wall function that is activated at that point). The other three computations predict a laminar separation bubble. Further downstream the skin friction increases, because of transition taking place. Predicting the laminar separation bubble seems to be linked to getting a transition at approximately the correct location. When there is no separation bubble, the reason is probably that the SGS model has been too dissipative and as a second effect the transition occurs too far downstream.

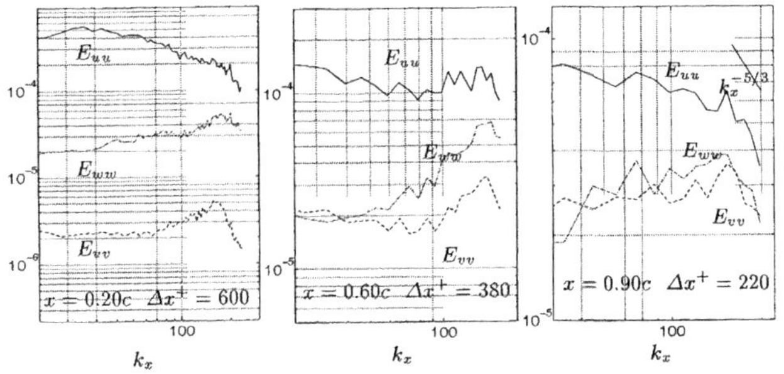

Figure 46 Chalmers: Approximate wave number spectra of the resolved velocities at $y/\delta \approx 0.12$ (from Smag3).

Extension in the Spanwise Direction. When periodic boundary conditions are used, these conditions imply that the extent in the spanwise direction should be wider than the largest structures in that direction. The largest scales in a boundary layer are in the order of the thickness (δ_{99}) and these scales are probably also relevant in the spanwise direction. Near the trailing edge the boundary layer thickness, δ_{99}, from the experiments, is approximately 9% of the chord. Table 18 shows the spanwise extent used in the different computations. Figure 44 shows the two-point correlation in the spanwise direction at three locations on the suction side. These correlations have been determined for Smag3 from the computation where the spanwise extent was 8% of the chord. This extent might not be sufficient. Keeping a fine resolution however is probably more valuable.

Turbulent Boundary Layer. A wall-resolved LES using a spanwise extent of 8% of the chord or more is too costly in terms of CPU time. If wall functions are used, in which the effect of the energy-producing structures near the wall is modeled (rather than resolved), grids with a coarser resolution can be used.

Table 20 Chalmers: Wall-boundary conditions used in the computations and the approximate averaged distance from the wall to the first node.

Computation	Wall-boundary condition	y_1^+
Smag1	no-slip	15
Smag2	no-slip	30
Smag3	wall functions, Eqs. 9-11 (Fig. 47)	40
Smag4	wall functions, Eqs. 9-11 (Fig. 47)	10
Yoshi1	wall functions, Eqs. 9-11 (Fig. 47)	10

However, the resolution still needs to be fine enough to capture the important large-scale turbulence. That is, the resolved turbulence should at least reach the inertial subrange if the fine-scale turbulence is to be represented with a SGS model. Energy spectra are shown in Fig. 45 at three locations on the suction side. These spectra are from the Smag3 computation and the resolved scales seem to reach inertial subranges except at 20% of the chord where the grid resolution is coarse ($\Delta z^+ = 350$). The resolution probably needs to be finer closer to the transition region. It is not clear why there are unphysical high-frequency oscillations in the energy spectra E_{ww}.

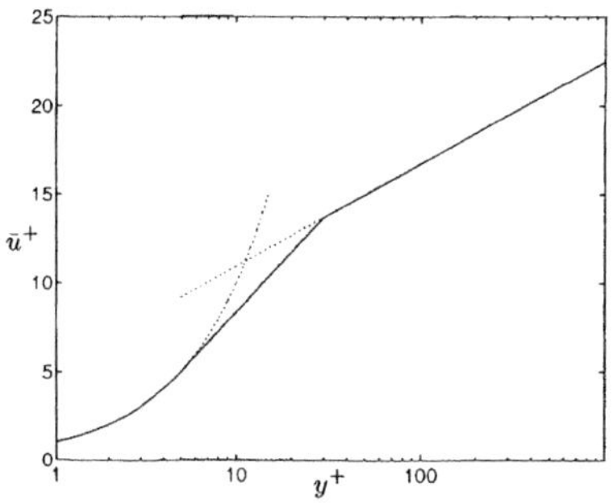

Figure 47 Chalmers: Plot of the wall functions (Eqs. 9-11).

Figure 46 shows approximate spectra in the streamwise direction using the assumption of 'frozen turbulence' (Taylor's hypothesis). The resolved velocities hardly seem to reach an inertial $-5/3$ range, except at 90% of the chord. The spectra indicate that the streamwise resolution is too coarse.

Perhaps a Δx^+ of approximately 200 is needed (resulting in e.g. a resolution that has to be three times finer at 20% of the chord).

Table 20 shows the wall-boundary conditions used in the turbulent region. The wall functions are based on the instantaneous log law in the log region ($y^+ \geq 30$):

$$\bar{u}^+ = \frac{\ln y^+}{\kappa} + B \,, \tag{9}$$

where $\kappa = 0.4$, $B = 5.2$. In the buffer region ($5 \leq y^+ \leq 30$) the following wall law is used:

$$\bar{u}^+ = \frac{\ln y^+}{C} + D \tag{10}$$

Eq. 10 is a matching between the log law (Eq. 9) and the linear law in the viscous sublayer ($y^+ \leq 5$):

$$\bar{u}^+ = y^+ \tag{11}$$

The constants in Eq. 10 become $C = \ln 6/(\frac{1}{\kappa}\ln 30 + B - 5)$ and $D = 5 - \ln 5/C$. For more details, see [50].

It is especially interesting to compare Smag2 with Smag3, since the wall functions are the only difference in the setup of these two computations. Using wall functions the skin friction coefficient is in better agreement with the experiments (Fig. 43), as well as the velocity profiles (Fig. 48). Also, the pressure coefficient does not bend down near the trailing edge (Fig. 54).

The wall functions do not have a large effect in the computations on the Common mesh (Smag4 and Yoshi1), where the first gridpoint adjacent to the wall is located at $y_1^+ \approx 10$ in order to be able to capture the thin separation bubble near the trailing edge which is observed in the experiments. The use of wall functions in this region is not feasible.

Figures 48-51 show resolved velocity and stress profiles at four locations on the suction side of the airfoil. In the computations on the Common mesh (Smag4 and Yoshi1), there are near-wall peaks in the stresses that result in fuller velocity profiles. Here $y_1^+ \approx 10$ and the wall functions have little effect.

At 96% of the chord the computation Smag1 and Smag4 have velocity profiles that are similar to the Smag3 computation close to the wall. However, in these computations (Smag1 and Smag4) the transition does not take place at the correct location. Also, in the Smag1 computation the resolution is much finer in the spanwise direction ($L_z = 0.03c$).

Lift and Drag Coefficients. Table 21 shows the lift and drag coefficients for the computations. The lift coefficients are overpredicted. Figure 55 shows the time history of the lift coefficient for computation Smag3 and Fig. 56 shows the frequency spectra of the signal. There is a peak in the frequency spectrum at a Strouhal number of approximately eight.

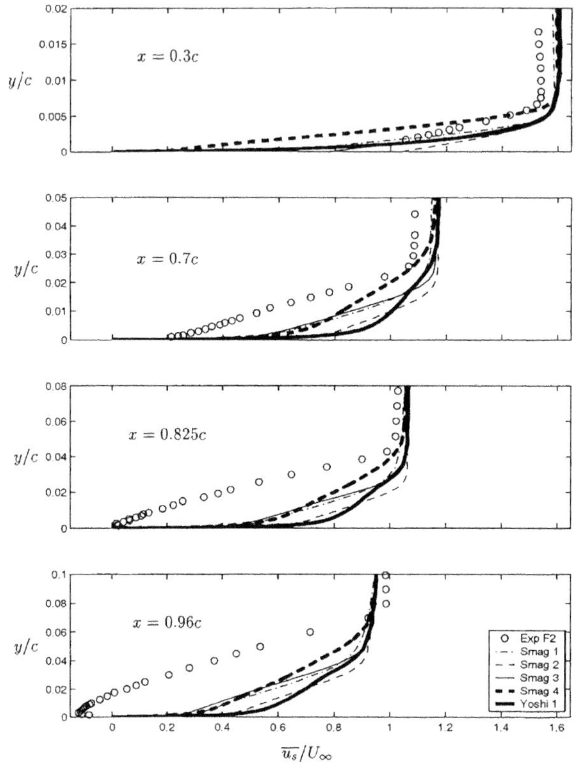

Figure 48 Chalmers: Spanwise and time averaged velocity in the streamwise direction.

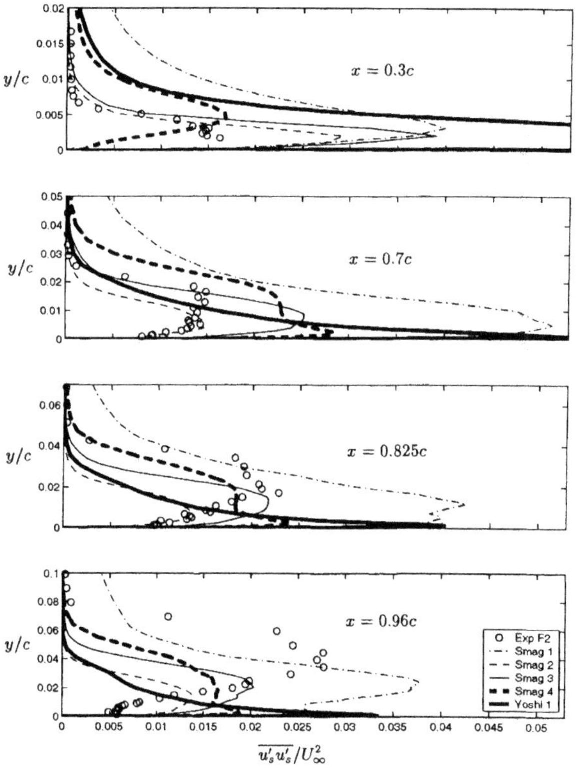

Figure 49 Chalmers : Spanwise and time averaged velocity fluctuations. Subscripts s and n denote the directions parallel and normal to the airfoil wall, respectively.

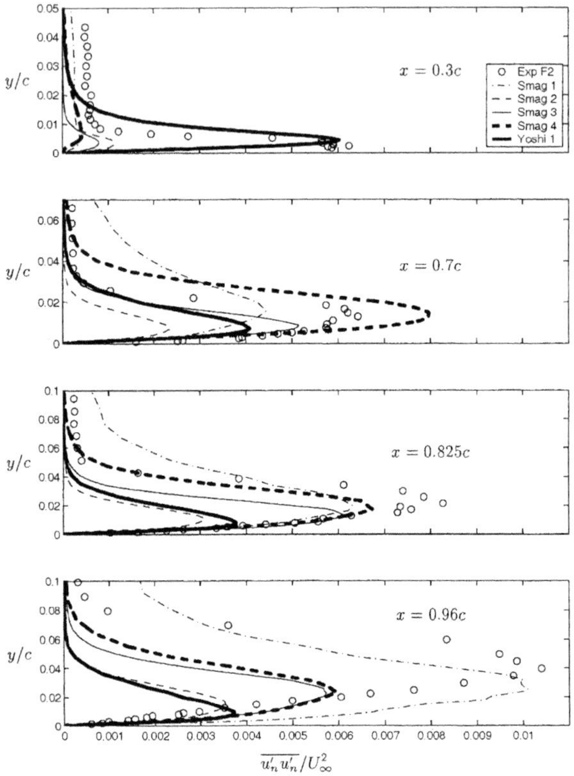

Figure 50 Chalmers : Spanwise and time averaged velocity fluctuations. Subscripts s and n denote the directions parallel and normal to the airfoil wall, respectively.

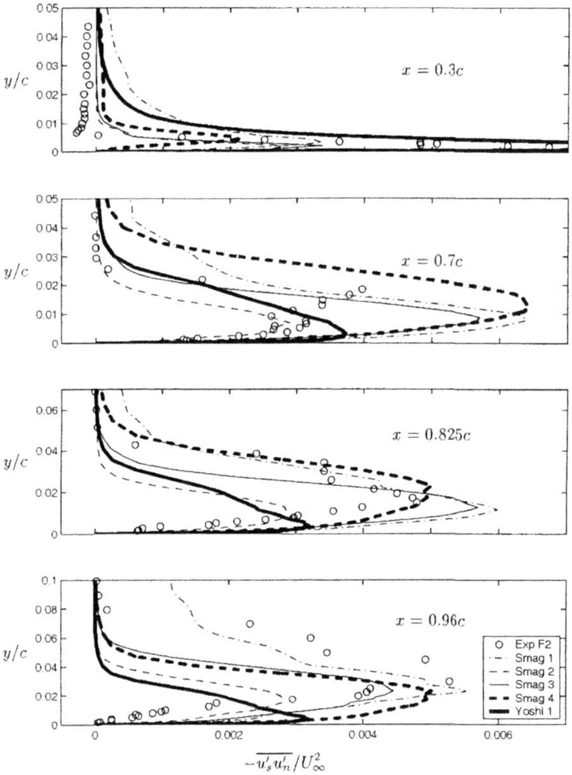

Figure 51 Chalmers : Spanwise and time averaged velocity fluctuations. Subscripts s and n denote the directions parallel and normal to the airfoil wall, respectively.

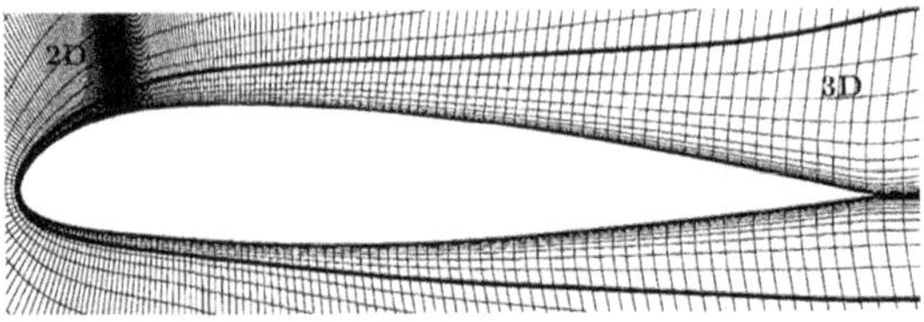

Figure 52 Chalmers: New mesh with increased resolution in the laminar and transitional region (1393 × 127 grid nodes). Every 4th node is shown.

Conclusions. In the computations undertaken meshes consisting of up to 1.6 million nodes have been used. The Chalmers and the Common mesh were constructed in order to keep the streamwise resolution below 600 wall units ($\Delta x^+ < 600$). It was hoped that this resolution would be sufficient when approximate wall-boundary conditions are used. However, this resolution is found to be insufficient in order to perform a satisfactory LES: energy spectra indicate that the grid resolution has to be approximately 200 wall units in the streamwise and spanwise direction in order to capture the large-scale structures *outside* the near-wall region.

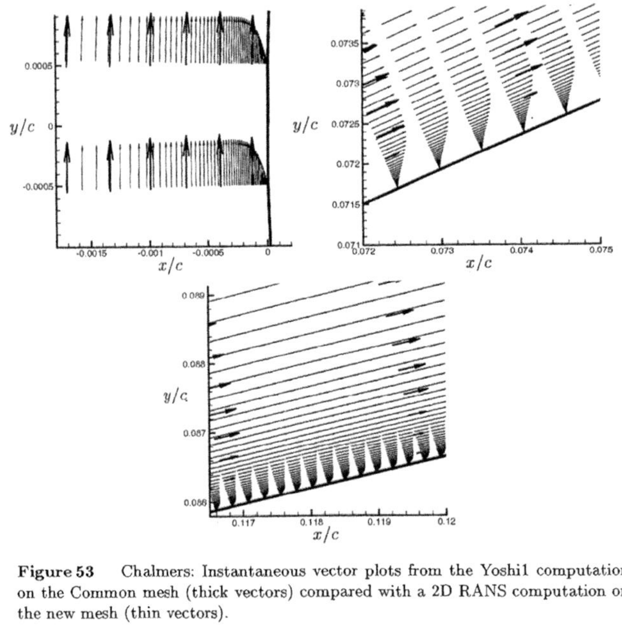

Figure 53 Chalmers: Instantaneous vector plots from the Yoshi1 computation on the Common mesh (thick vectors) compared with a 2D RANS computation on the new mesh (thin vectors).

The near-wall region has been modelled using wall functions based on an instantaneous wall law. Compared to when no wall functions are used, the results are improved. Since wall-functions have little or no effect when used with the relatively fine Common mesh, the near-wall region needs to be modelled using some other approach (e.g. RANS).

Second-order upwinding is used in parts of the domain to suppress odd-even oscillations. These wiggles are found mainly in the area around the leading edge. Unless they are suppressed, unphysical oscillations contaminate the entire computational domain.

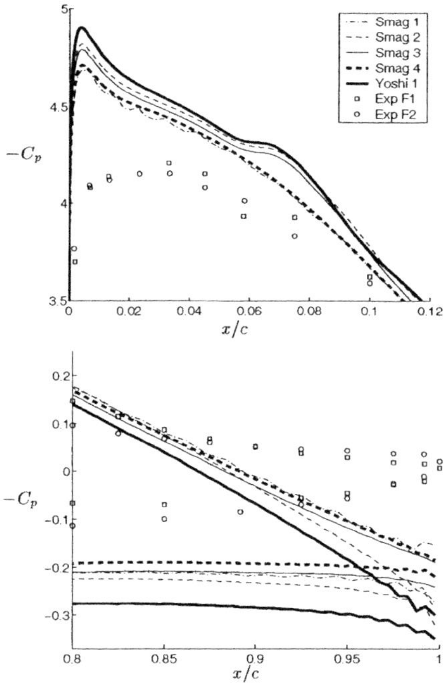

Figure 54 Chalmers: The pressure coefficient at the leading and trailing edge.

Also, the second-order upwinding in the mixed scheme prevents any numerical transition from taking place. At approximately the location where the CDS is fully active the transition takes place, *provided* that the SGS model is not too dissipative in this region. In the computations where no model is used upstream of the nominal transition point there are peaks in the resolved stresses at the correct location. In these computations (with zero eddy viscosity in the laminar region) incipient laminar separation bubbles are predicted. However, the laminar boundary layer develops incorrectly because of the poor resolution and the incipient bubbles are located too far upstream (at approximately 7% of the chord).

Table 21 Chalmers: Spanwise and time-averaged lift and drag coefficients for the computations and the experiments.

	Exp. F1	Exp. F2	Smag1	Smag2	Smag3	Smag4	Yoshi1
C_L	1.574	1.515	1.72	1.75	1.72	1.70	1.78
C_D	0.0212	0.0308	0.0280	0.0167	0.0170	0.0233	0.0155

Because of the strong coupling between the trailing edge separation and the pressure peak at the leading edge, all regions around the airfoil are probably equally important. In order to impose the transition in a proper way, perhaps a minimum requirement is to resolve the laminar boundary layer and the laminar separation bubble. With a mesh (Fig. 52) and a 2D/3D approach the number of nodes becomes 3.2 million. In on-going work a hybrid LES-RANS is used, with RANS in the very near-wall region coupled with LES in the outer region [52].

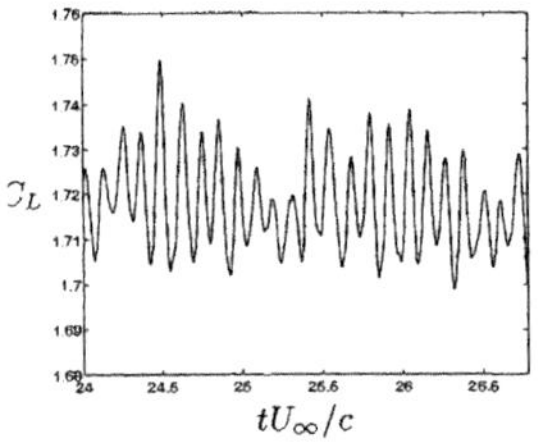

Figure 55 Chalmers: Time history of the lift coefficient C_L for computation Smag3.

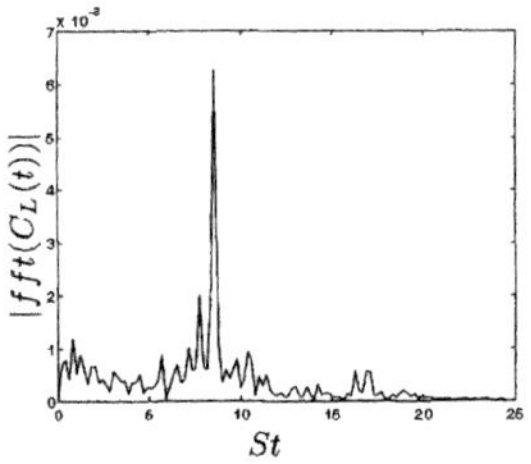

Figure 56 Chalmers: Frequency spectrum of the C_L signal from computation Smag3.

4.3 Contribution by Alenia

Author: N. Ceresola

Objectives. The purpose of the Alenia participation in the LESFOIL project was to acquire knowledge on the application of LES techniques to external flow of industrial interest and to make a first assessment of the validity and efficiency of using LES in such cases. To allow a comparison with more traditional and computationally cheap methods, Alenia computed the common testcase (A-airfoil) using the simulation tools of use for industrial computations.

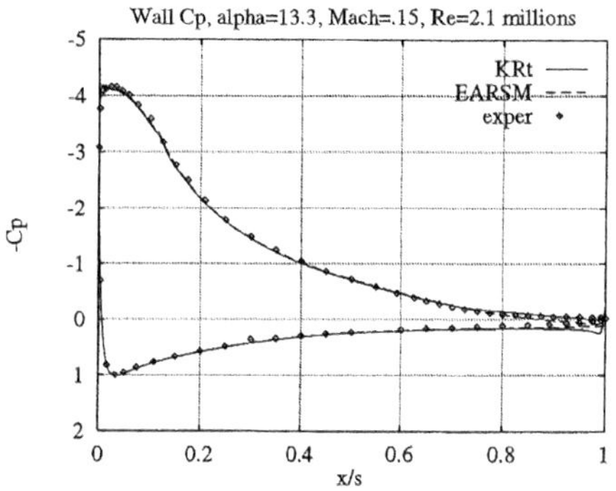

Figure 57 Alenia: Wall C_p

Numerical method. The computations have been performed using the code UNS3D. The solution algorithm is based on a finite volume, node centred approach operating on an unstructured grid. For each grid node, a dual cell C_i is defined on which the flux balance has to be calculated. This is accomplished by splitting the volume boundary into edges ∂C_i, so that each edge is shared by a pair of cells C_i and C_j . The corresponding numerical flux is a function of the flow variables at C_i and C_j and of the integrated normal to the corresponding edge. The flux balance reduces then to the sum of the contributions coming from each edge, on which the fluxes are assumed to be constant. The overall sum of the contributions of the cell edges to the surface integration can therefore be made on a segment basis. Given the connectivity rules between the segments of the grid, the solver can deal therefore without distinction with tetrahedra, hexahedra and prisms, even mixed to form hybrid grids.

The artificial dissipation model is derived from the nonlinear scheme of Jameson. The equations are integrated in time with a second order backward difference and the dual time stepping technique. A five stage Runge-Kutta scheme is used to drive the residual toward zero at each timestep. In practice, it was considered to be sufficient to reduce the residual by two orders of magnitude at each timestep. With the use of implicit residual averaging, a local CFL number of 4 could be employed in the five stage multistage subiteration process.

The Weiss and Smith version of low Mach number preconditioning is implemented in the code. For the computation of the A-airfoil testcase, its application was found to be necessary in order to find a converged solution in a reasonable number of iterations.

Turbulence models. Two turbulence models have been applied:

1. The two-equation $k - R_t$ model, where R_t is the turbulence Reynolds number $k^2/(\nu\epsilon)$. The Shear Stress Transport (SST) limiter is employed in the computation of eddy viscosity, to limit the turbulence production in regions where it exceeds dissipation [53]. One characteristic of the model is that it does not employ the wall-distance, thus allowing a more flexible application to complex geometries on unstructured grids.
2. The Wallin-Johansson Explicit Algebraic Reynolds Stress Model (EARSM) [54]. In this formulation, an explicit expression of the Reynolds stresses is obtained without imposing any condition to the production-to-dissipation rate. The isotropic part of the Reynolds stresses is expressed as a nonlinear eddy viscosity, which enforces the numerical stability of the method.

A point implicit treatment of the source terms is implemented, which has demonstrated to improve the convergence properties of the calculations.

Steady flow computations. Computations have been carried out on the A-airfoil testcase at 13.5 degrees of incidence, $Ma = 0.15$ and a Reynolds number of 2.1 million, using turbulence model of industrial use. The computational grid which was used was the ONERA 513 x 97 Gold mesh.

In the first computation, the $k - R_t$-SST model with a two-equation turbulence model was used. The second computation was made applying the Explicit Algebraic Reynolds Stress (EARSM) model of Wallin and Johansson, using the $k - R_t$ equations to obtain the turbulence quantities.

Due to the presence of a laminar separation around $x/c = 0.12$, followed by a transitional bubble, fixing transition was of crucial importance for our computations. In the present calculations, transition was triggered at $x/c =$

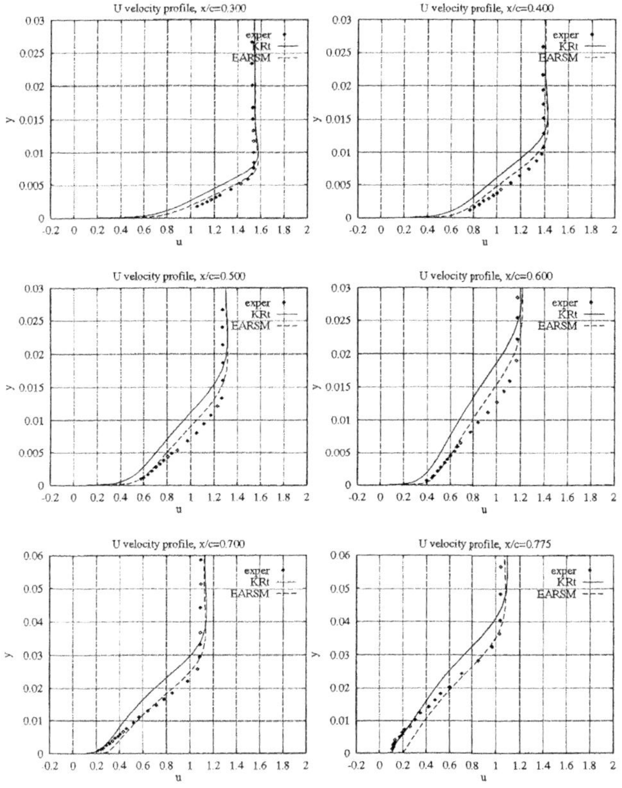

Figure 58 Alenia: Streamwise U profiles

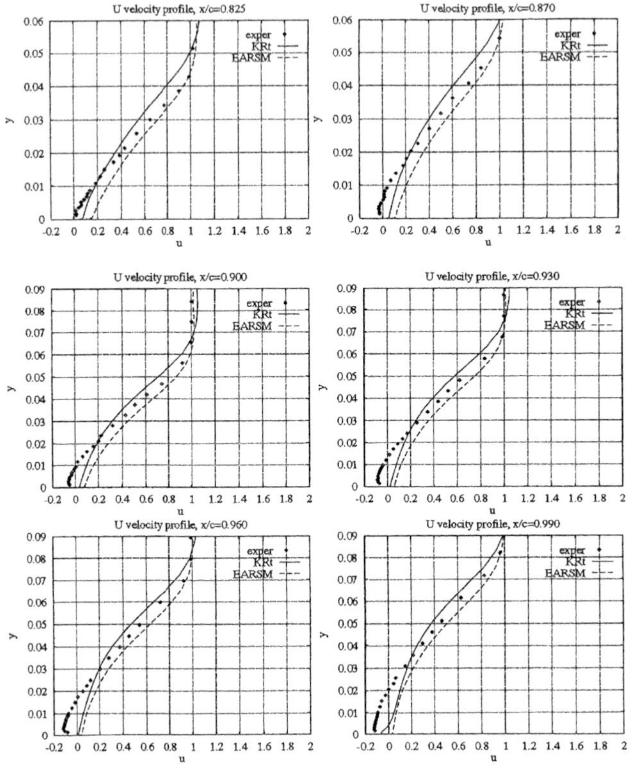

Figure 59 Alenia: Streamwise U profiles

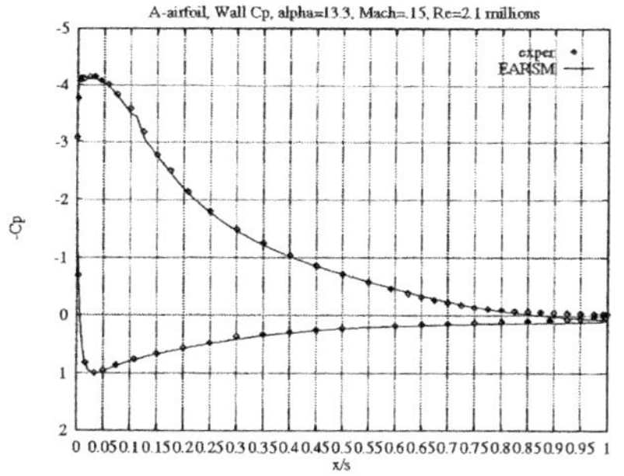

Figure 60 Alenia: URANS: Wall C_p

0.12. With EARSM model a sudden increase of turbulence production was found after that location, while for RANS it was necessary to introduce a "numerical transition strip" (following Spalart and Allmaras, [55]) to obtain transition at the correct location. This may have affected the velocity profiles at the initial stations.

Results. The obtained wall C_p distributions are shown in Fig. 57. Here the main difference between the two solutions appears to reside in the region near the trailing edge, where a small separation is present in the experiment. The $k - R_t$-SST model seems to better predict the small acceleration of the flow on the upper side, due to the presence of the recirculation bubble. An explanation of what is happening can be devised looking at the U velocity profiles in Figs. 58 and 59. After $x/c = 0.775$, a more unstable velocity profile is predicted by the $k - R_t$-SST model according with the experiment, while stable boundary layer profiles are obtained with EARSM. Neither model is able to predict the presence of the recirculation region after $x/c = 0.87$, though a very thin bubble can be seen at the last station with the $k - R_t$-SST model.

Unsteady RANS computations. To investigate the effect of a possible unsteadiness of the flow at trailing edge, it was decided to do unsteady computations on the same testcase. The Explicit Algebraic Reynolds Stress Model was employed, coupled with the $k - R_t$ model. Transition was triggered introducing a "numerical transition strip" (following Spalart and Allmaras [55]) to obtain transition at the correct location.

The steady state flowfield was used to initialize the computation, which was initially carried out with a reduced Δt of 5E-03. A Δt of 5E-04 was then employed, to allow for a more significant comparison with LES results.

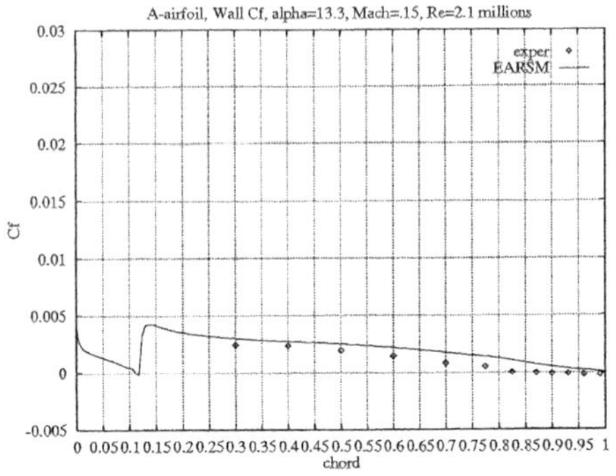

Figure 61 Alenia: URANS: Wall C_f

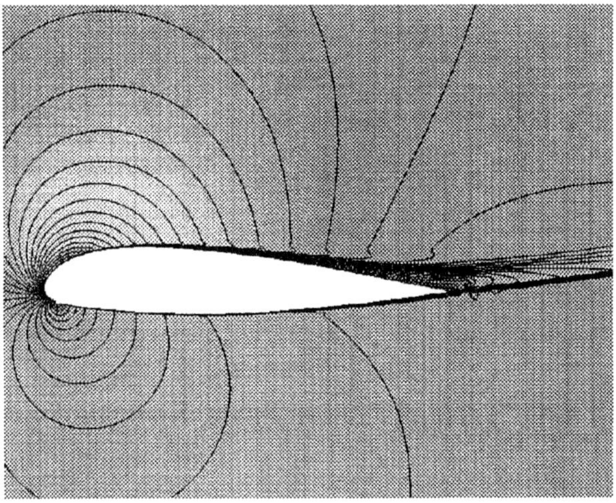

Figure 62 Alenia: URANS: Instantaneous Mach number levels

Results. No significant changes in C_p (Fig. 60) and Cf (Fig. 61) were found, with respect to the steady state result.

Nevertheless, unsteadiness was found to be present at the trailing edge, due to an oscillation of the vortex sheet stemming from the sharp trailing edge. As shown in Fig. 63, the vertical component of velocity is oscillating, reversing periodically its sign, with a Strouhal number of about 8. An interesting fact is that a sharp peak in the lift coefficient was found by Chalmers using LES at the same frequency.

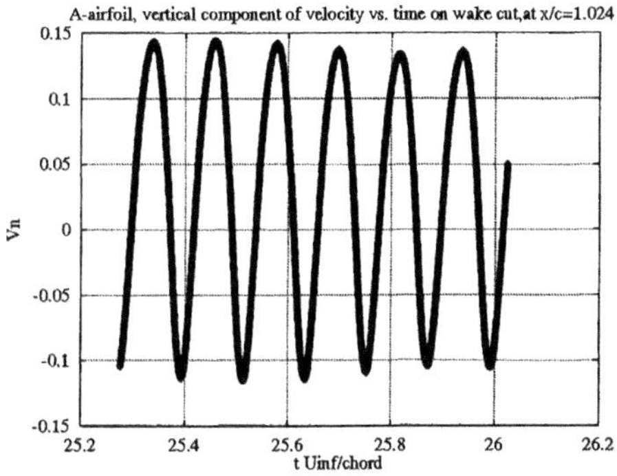

Figure 63 Alenia: URANS: Vertical component of velocity vs. time on wake cut, $x/c = 1.024$

A qualitative picture of the oscillating sheet is depicted in Figs. 62 and 64.

Conclusions. The computations on Task 5 testcase have confirmed that the RANS turbulence models which have been employed can correctly simulate the attached boundary layer past the airfoil, while they fail to predict the thin recirculation region taking place near the trailing edge. Even the computation in unsteady mode did not show appreciable improvement with respect to the steady state simulation. The question remains however, if the inclusion in the geometrical model of a blunt trailing edge as in the experiment would have improved the results of the simulation.

Recommendations for future work. From an industrial point of view, the best compromise between efficiency and accuracy would be to retain the less expensive RANS models where we know they are performing well, that is for most applications of engineering interest. LES simulation, on the other hand, may be necessary where large, unsteady structures are present in the flow. The new approaches based on hybrid LES-RANS simulation may be therefore worth to be investigated and developed in the future.

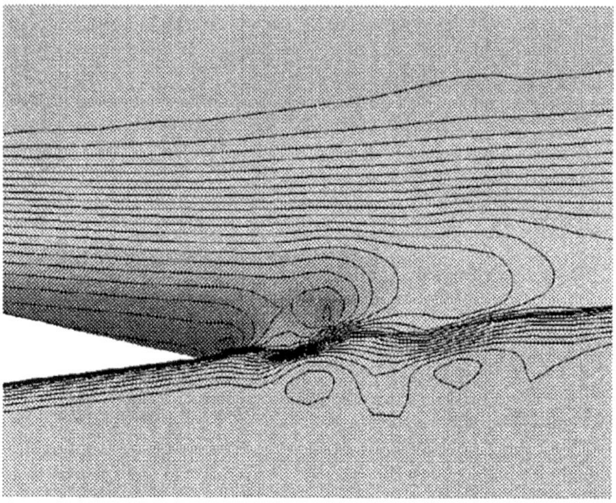

Figure 64 Alenia: URANS: Instantaneous Mach number levels

4.4 Contribution by CERFACS

Author: Frédéric Ducros

Introduction. The simulations presented in this report were performed using the CERFACS (Silicon Graphics Origin 2000) and Météo France facilities (VPP 700 & 5000). The codes employed were the structured NSMB code, currently used by Aérospatiale and the unstructured platform AVBP, developed by CERFACS and Oxford University. Both solve the complete compressible Navier-Stokes equations, include both LES and RANS turbulence modelling, and are parallelised.

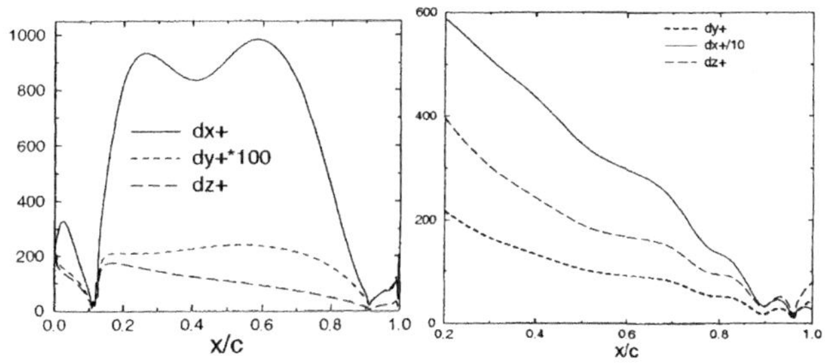

Figure 65 CERFACS: Resolution at the wall for meshes M1 (**left**, see also [56]) and M3 (**right**). The friction velocity is taken from Simu 1 and 3 respectively and starts at $x/c = 0.2$ for the WF mesh because u_{tau} is not representative before $x/c = 0.2$ (see the text).

After outlining the basic concept of the Wiggle Detector and some of the features of the use of wall functions at high Reynolds number this contribution describes the results obtained from three different computations of the A-profile: the first completes the pioneering LES of Weber and Ducros [57], [58], [56] using the filtered structure function model [59]; the second one focuses on the use of wall functions; the last one on the effect of the shape of the trailing edge. For these two last simulations we made use of the WALE SGS model [60].

An aborted attempt at using the Detached Eddy Simulation (DES) promoted by Spalart [24] is briefly discussed here. In this computation the initial flow field was from a steady state solution obtained using the Spalart-Allmaras (SA) model and the ONERA gold mesh. The DES was then conducted using an implicit time step ($\Delta t = 0.0018c/U_\infty$) larger than the one used

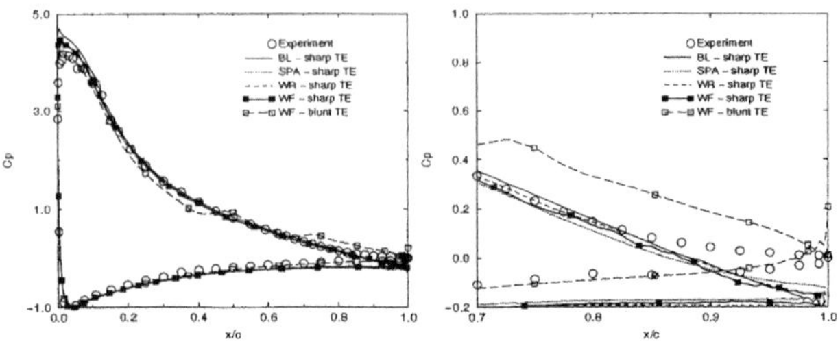

Figure 66 CERFACS: Pressure coefficients around the airfoil for Simu 1 to 3. Steady RANS calculations obtained using the M1 mesh (see Table 22) are also reported: BL and SPA indicate the Baldwin-Lomax and the Spalart-Allmaras models respectively.

for the LES The spanwise extent of the domain was selected to be $L_z = 0.08c$ with 20 computational points. After one advection time (c/U_∞), the flow was nearly unchanged, exhibiting only very weak 3D structures. In consequence, the statistics followed the SA results and are not reported here. Certainly, better results would be expected when using a smaller time step and/or a different mesh which alters the critical position of the switch between the RANS and LES modelling. However, this will probably lead to the use of meshes and grid spacings similar to those usually adopted for standard LES. This is not what is expected from DES. As a result DES has not been intensively investigated, even if it is an approach which is worth looking at [61], [62].

Numerical schemes. The numerical schemes used are all based on the Jameson scheme, either on its original form, or on a high-order form [63]. These schemes are basically centered and so do not introduce any diffusion which would be able to regularize wiggles which may develop due to both compressibility and distorted meshes. A remedy exists in the use of a 'wiggle-detector' to localize the wiggles and the introduction of a limited amount of artificial dissipation which damps the wiggles. The principle of the so-called 'wiggle detector' is explained here using the basis of a structured code; similar developments have been undertaken for the unstructured code (AVBP) and have led to similar effects.

Consider point (or cell) i. This point (cell) and the neighbouring points (cells) $i-2$, $i-1$, $i+1$ and $i+2$ form the stencil of the scheme on the x-axis. We consider now the plane defined by the axes (x, W) where W is a scalar quantity possibly exhibiting wiggles (here it is the energy ρe). We denote by M_i the point of the coordinates $(x_i, W_i, 0)$. The vectorial product $Pvi = M_{i-1}M_i \times M_iM_{i+1}$ is then a vector parallel to the z-axis. It is given by:

$$\begin{aligned} Pvi &= (0, 0, P_i) \\ P_i &= M_{i-1}M_i \times M_iM_{i+1} \\ &= \Delta_{i-1/2}(W_{i+1} - W_i) - \Delta_{i+1/2}(W_i - W_{i-1}) \end{aligned} \tag{12}$$

where $\Delta_{i+1/2}$ is the distance between computational points i and $i+1$ in the case of a finite difference formulation, or the distance between the centers of computational cells i and $i+1$ if the finite volume approach is considered. The component P_i has the same sign as the discretized second-order derivative of W at point (cell) i. We consider that a wiggle is present in the set of points (or cells) $i-2$ to $i+2$ if the sign of the second-order derivative changes twice over this interval. This can be expressed by the following condition:

$$\begin{array}{lll} S_{i,-} + S_{i,+} = -2 & & \text{with} \\ S_{i,-} = \psi_{i-1}\psi_i \quad S_{i,+} = \psi_i\psi_{i+1} & & \\ \psi_i = +1 & & \text{if} \quad P_i > 0 \\ \psi_i = 0 & & \text{if} \quad P_i = 0 \\ \psi_i = -1 & & \text{if} \quad P_i < 0 \end{array} \tag{13}$$

If the condition expressed by Eq. 13 is satisfied, a dissipative term is added to the physical fluxes, otherwise it is omitted. This results in a centered, stable scheme.

Wall functions. Two different wall function models were tested. The Werner and Wengle [18] law of the wall has been implemented in the NSMB solver using a finite volume approach based on cell centered collocated variables. A more classical but implicit wall function based on a two-layer approach has been used for the unstructured code. These laws have been validated in the case of a periodic channel flow.

Airfoil Calculations

Mesh. As previously mentioned, three simulations have been carried out, hereafter referred to as Simu 1 to 3: the corresponding meshes are noted M1 to M3.

The first simulation adopted the mesh used by Weber for his first LES [56]. Weber's simulation has been continued during the project using the WALE model, but no noticeable change was observed in the flow profiles. Hence, the results of Simu 1 are identical to those in [56] and are taken over from this reference. However, some differences were noticed in the eddy viscosity fields. The M1 mesh was also used for RANS calculations with the Baldwin-Lomax (BL) and Spalart-Allmaras (SPA) model.

Table 22 CERFACS: Information on the simulations presented: Wall BC stands for wall boundary conditions. WR = wall resolving, WF = wall function, FSF = filtered structure function, WALE = Wall Adapting Local Eddy Viscosity.

Simu	type of mesh	points	Wall BC	L_z/c	SGS model
1	M1 structured	498 × 64 × 30	WR	0.036 c	FSF and WALE
2	M2 structured	256 × 64 × 18	WF	0.020 c	WALE
3	M3 unstructured	8409 × 25	WF	0.036 c	WALE

The meshes used in simulations 2 and 3 were provided by ONERA (structured: sharp trailing edge, see Fig. 1) and Dassault-Aviation (unstructured: blunt trailing edge, see Figs. 76 and 77) respectively. The last mesh consists of 399288 cells. The corresponding wall surface resolution in terms of wall units is provided in Fig. 65.

Transition modelling. The experimental data show a transition at $x/c = 12\%$ on the upper side while the transition was tripped at $x/c = 30\%$ at the lower side. From a numerical point of view the adopted resolution prevents a realistic simulation of transition (see Table 22). Consequently numerical tricks were employed so as to force the transition: white noise has been added on the pressure side at $x/c = 30\%$ (amplitude $10^{-3}U_\infty$ for Simu 1 and 3, $10^{-2}U_\infty$ for Simu 2). For Simu 1, the transition was forced by relaxation of the artificial dissipation used to suppress the wiggles generated at the stagnation point (the method used by Weber and Ducros ([56])). An alternative simulation has been performed by removing this artificial damping and letting the flow undergo transition spontaneously. This has been possible by using the wiggle detector previously mentioned. In both cases, the results are similar, even in the regions close to the leading edge. For Simu 2 and 3 (using wall functions), the first calculations permitted the wall function models to act by themselves. In both cases, although the model should logically switch by itself from the laminar to the turbulent state and thus prescribe the correct friction on the wall, it was seen that often a turbulent state was identified by the model in pre-transition regions on the suction side (for $x/c \leq 12\%$), which is incorrect. In the case of Simu 3, the first results were particularly bad since the excessive friction near the leading edge led to a too large thickening of the boundary layer and produced an incorrect prediction of the pressure coefficient. A basically unsatisfactory but "efficient" remedy has been used. This is:

- For Simu 2, a laminar state is prescribed for the model between $x/c = 0$ and $x/c = 0.12$. The model is switched on for larger x/c.
- For Simu 3, no friction is applied between $x/c = 0$ and $x/c = 0.12$. This explains why no friction estimate is supplied before $x/c = 0.2$. We know this is unsatisfactory but so far no better solution has been found. The model is switched on for larger x/c.

– In the RANS calculations reported using the BL and SA models, the eddy viscosity was set to zero in the laminar region and the turbulence models were activated at the experimentally observed transition stations.

Table 23 CERFACS: Information on how transition is treated in simulations 1 to 3. Note: WN stands for white noise.

Simu	Transition on the lower side	Transition on the upper side
1	triggered at $x/c = 30\%$	artificial diffusion up to $x/c = 12\%$
2	triggered at $x/c = 30\%$	triggered by WN at $x/c = 12\%$
3	triggered at $x/c = 30\%$	triggered through WF application

Results. Establishing a hierarchy between the following results is difficult, yet Simu 2 is considered to be the most satisfactory simulation, because (1) it provides lift and drag coefficients which are closer to those of the experiment, and (2) it provides results as good as Simu 1 while being less expensive. The selection of Simu 2 is made despite the fact that no separation is predicted by this computation.

Table 24 CERFACS: Time integration method and time step for the three simulations. DTS indicates Dual Time Stepping, RK4 Runge Kutta of 4th order (cfl = 1.5) and LW the Lax-Wendroff method (CFL = 0.6)

	Simu 1	Simu 2	Simu 3
numerical method	DTS	RK4	LW
$\Delta t U_\infty / c$	$1.8\ 10^{-4}$	$2.96\ 10^{-6}$	$9.2\ 10^{-7}$

Aerodynamic coefficients. The results of Simu 1 were interpolated onto the meshes of Simu 2 and 3 and taken as initial conditions. The same conditions have been adopted by Dassault-Aviation for their simulation. Table 24 provides the integration time step for all simulations, showing the clear difference between the dual time stepping method (DTS) and the explicit ones. The time step chosen for the DTS method results from preparatory studies ([58] and [57]) and corresponds to approximately $\delta/U_\infty/10$, δ being the boundary layer thickness.

Table 25 provides the mean values for the lift and drag coefficients. The Simu 3 results, though they exhibit better mean streamwise velocity profiles (given later) and a C_p closer to experimental values near the leading edge, return a poorer estimation of the global C_L. The overall tendency is that, as

Table 25 CERFACS: Lift and drag coefficients obtained by Simu 1 to 3 compared with experimental data. C_D is not provided for Simu 3 because the method used to switch on the wall function perturbs the global C_D too much. BL = Baldwin-Lomax model and SPA = Spalart-Allmaras model (both RANS).

	BL	SA	Simu 1	Simu 2	Simu 3	Exp. F1
C_L	1.72	1.64	1.68	1.655	1.75	1.56
$C_{Lrms}(\%)$	0.	0.	6	0.7	30	/
C_D	0.0217	0.0220	0.0230	0.022	/	0.0204

far as C_L is concerned, Simu 3 results correspond to the more diffusive RANS results. For all the simulations which used the sharp trailing edge the mean aerodynamic coefficients exhibited the same type of time evolution, whereas larger variations are found for the blunt trailing edge. This corresponds to the dynamics of the flow for each simulation and shows that the unsteadiness of the flow is more pronounced for Simu 3 than for the other simulations. These features are confirmed by the analysis of the spatial distribution of C_p and C_f (see Fig.'s 66 and 67).

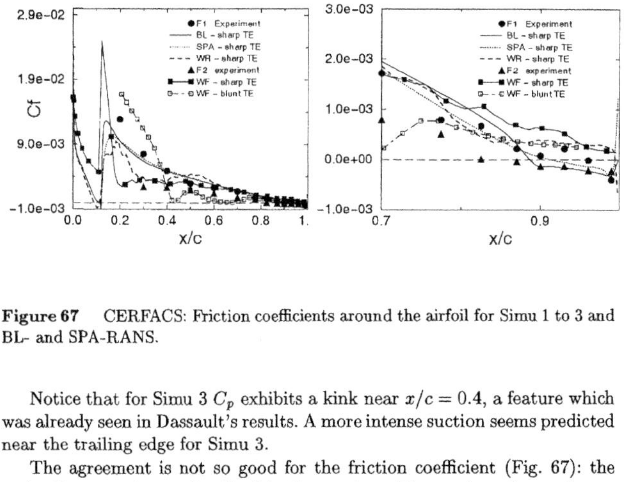

Figure 67 CERFACS: Friction coefficients around the airfoil for Simu 1 to 3 and BL- and SPA-RANS.

Notice that for Simu 3 C_p exhibits a kink near $x/c = 0.4$, a feature which was already seen in Dassault's results. A more intense suction seems predicted near the trailing edge for Simu 3.

The agreement is not so good for the friction coefficient (Fig. 67): the main discrepancies are localized in the pre-transition region where a larger friction coefficient is found. A very high peak of friction coefficient occurs at the imposed transition location on the suction side. The relaxation of the friction coefficient is then too large, leading to an underestimation of C_f up to $x/c \approx 50\%$.

Velocity profiles. Only the structured meshes are fine enough in the near wake region to provide reasonable resolution in this region, see Figs. 1 76 and 77. The results are that both Simu 1 and 2 provide the same type of results, under-predicting the velocity defect. Considering the profiles on the airfoil (Fig. 69), the results appear reasonable for $x/c = 0.4$, even if they do not indicate correct turbulence profiles. At $x/c = 0.7$, Simu 1 and 2 predict profiles typical of fully attached flow whereas the experimental profiles already exhibit some kind of inflexion. Simu 3 produces a too pronounced inflexion profile. For the remaining stations, the result of Simu 3 seems closer to the experiments than the other simulations (Figs. 68 and 69).

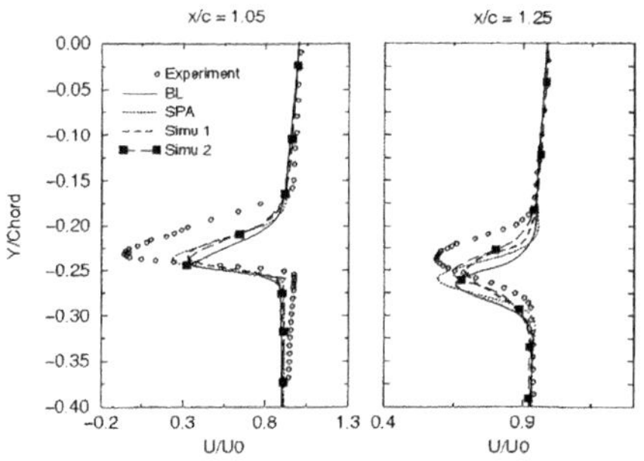

Figure 68 CERFACS: Mean streamwise velocity in the wake for Simu 1 and 2. and the BL- and SPA-RANS calculations.

This hierarchy observed between Simu 1, 2 and 3 is not the same when looking at the turbulent shear stress (Fig. 70) and streamwise velocity fluctuations (see Fig. 71). These profiles confirm a known feature of LES results, namely a too high level of resolved fluctuations arising as a result of low resolution – this trend seems worse when wall functions are used. On the other hand it appears that the dynamics are not the same for Simu 2 and 3. Larger unsteady structures seem to appear for Simu 3. These trends are confirmed by the visualizations.

Velocity contours. Figure 72 provides contours of the streamwise velocity for some of the computations. The main conclusions are:

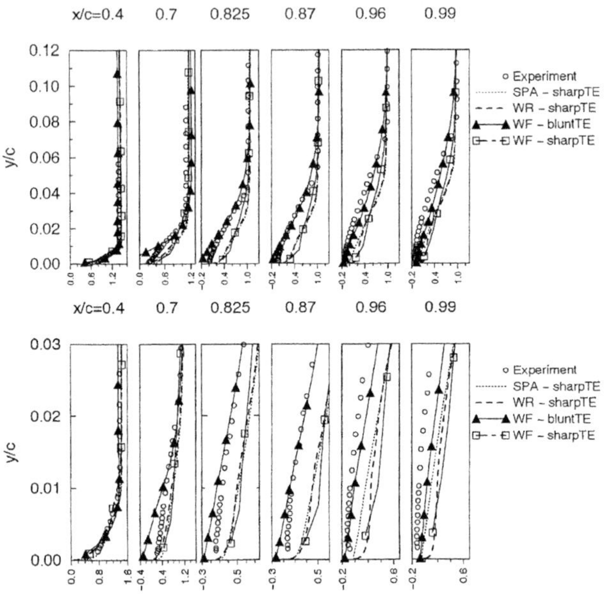

Figure 69 CERFACS: Mean streamwise velocity on the suction side for Simu 1, 2, and 3 and for RANS calculations with the SPA model: global (**Top**) and near-wall close-up (**Bottom**) views. The abscissa is U/U_0.

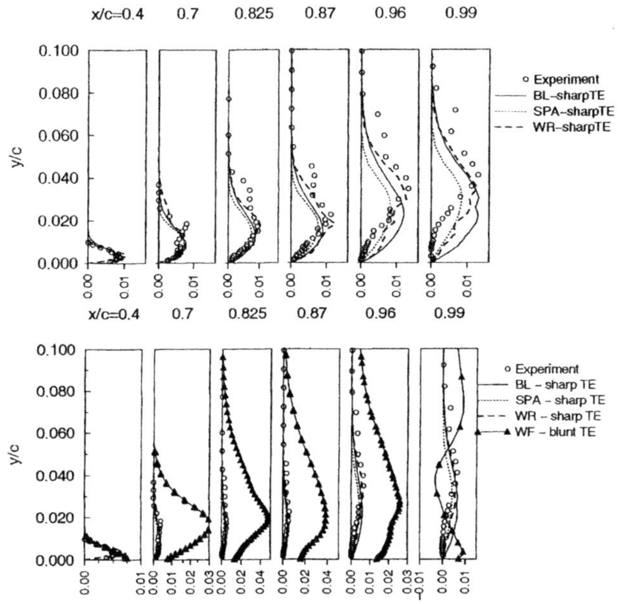

Figure 70 CERFACS: Turbulent shear stress $(-\overline{u'v'}/U_0^2)$ on the suction side for Simu 1 and 3 including BL- and SPA-RANS calculations. The top figure provides an enlarged view of the results from Simu 1 (see also [56]) From [56] with permission by Taylor & Francis Ltd. http://www.tandf.co.uk/journals.

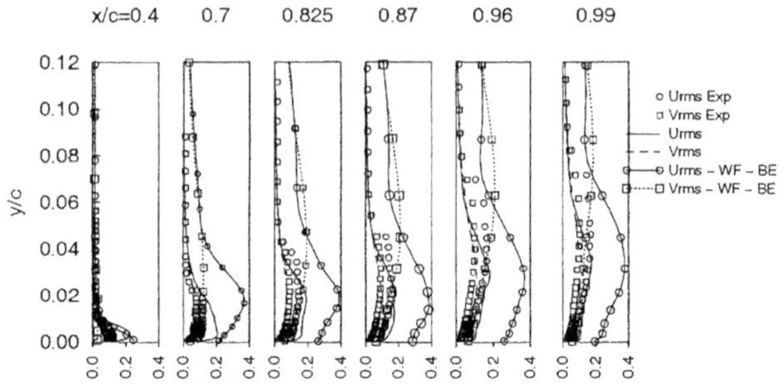

Figure 71 CERFACS: Resolved velocity fluctuations $\langle u'u' \rangle^{1/2}/U_0$ and $\langle v'v' \rangle^{1/2}/U_0$ on the suction side for Simu 1 and 3.

- Averaged LES simulations produce solutions very close to the equivalent simulations provided by RANS approaches (using Baldwin Lomax, Spalart Allmaras or $k - \epsilon$ models).
- The larger separation found for the blunt trailing edge is clearly visible from both the instantaneous and the averaged solution of Simu 3.
- Simu 3 produces in fact a too large separation. This result is already visible in the mean streamwise velocity profiles.

Eddy viscosity behaviour. Figure 73 displays the mean eddy-viscosity profiles for Simu 1 and 3 and for the RANS simulations. The ratio of turbulent to molecular viscosity, ν_t/ν is plotted in the figure. Except for the transition area on the suction side the larger values come from the RANS calculations, as expected. The LES approaches yield the largest eddy viscosity near the transition region, namely at $x/c \approx 0.2$ to 0.4. The eddy-viscosity level is nearly one order of magnitude larger for the wall function mesh than for the wall resolved simulation. This is due to both the size of the cells (resulting in an increased filter length scale) and the higher intensity of the velocity fluctuations (see Fig. 70). A cut of the instantaneous eddy-viscosity is shown in Fig. 74 for Simu 3 with the blunt trailing edge. Very large scales detached from the airfoil are visible: they were not identified for the simulations with the sharp trailing edge. This phenomenon corresponds to the highly detached flow found with Simu 3.

Spanwise averaged distributions of eddy-viscosity obtained for the same flow field with two different SGS models are shown in Fig. 75. Both models are almost inactive in regions of laminar flow, in particular around the leading edge and are able to naturally switch on in the turbulent flow regions where their action is necessary. The Wall-Adapted Filtered Structure

Figure 72 CERFACS: Contours of streamwise velocity at a given spanwise cut for Simu 3 (**top left:** average; **top right:** instantaneous, see also [56]) and for Simu 1 (**middle left:** average; **middle right:** instantaneous) and for a steady SPA-RANS computation (**bottom**). From [56] with permission by Taylor & Francis Ltd. http://www.tandf.co.uk/journals.

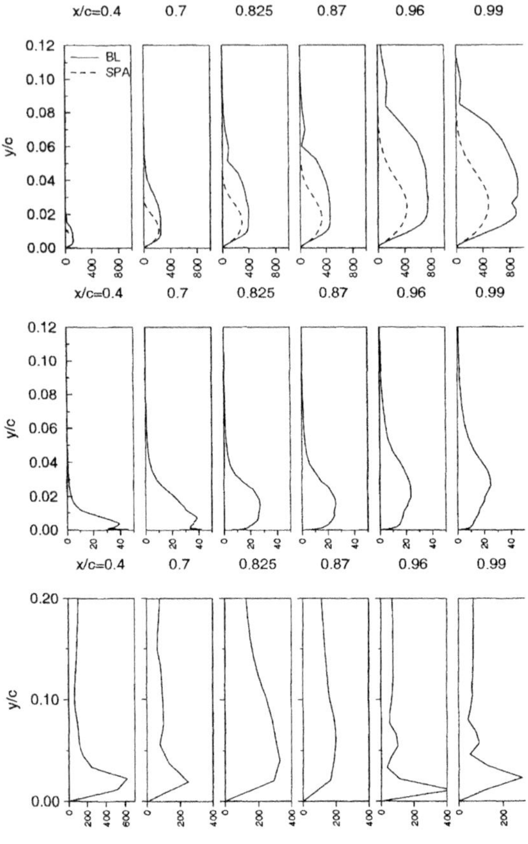

Figure 73 CERFACS: Mean eddy-viscosity profiles for Simu 1 (**Middle**, see also [56]) and 3 (**Bottom**), as well as reference eddy viscosity for the RANS simulations (**Top**) with the BL- and SPA models. The ratio ν_t/ν is shown. From [56] with permission by Taylor & Francis Ltd. http://www.tandf.co.uk/journals.

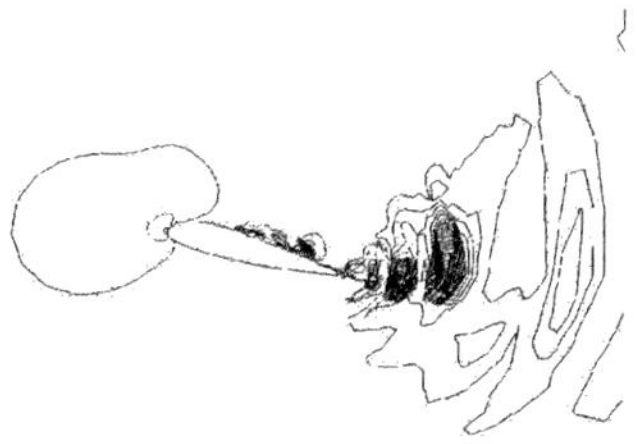

Figure 74 CERFACS: Eddy-diffusivity field in one spanwise cut for Simu 3. Large separation zones are identified by concentrated isolines, showing large detached structures.

Function (WAFSF) model exhibits a smoother distribution of eddy-viscosity than that of the WALE model. This has been identified as being due to an increased sensitivity of the model to wall-normal small structures. As in the channel flow both models exhibit a y^3 asymptotic behaviour of the eddy-viscosity close to the wall (Fig. 75). Also, consistent with the channel test-case results and with the values of the model constants ($C_{WALE} = 0.1$ and $C_{WAFSFS} = 0.043$), the structure function model provides an eddy-viscosity of slightly higher level.

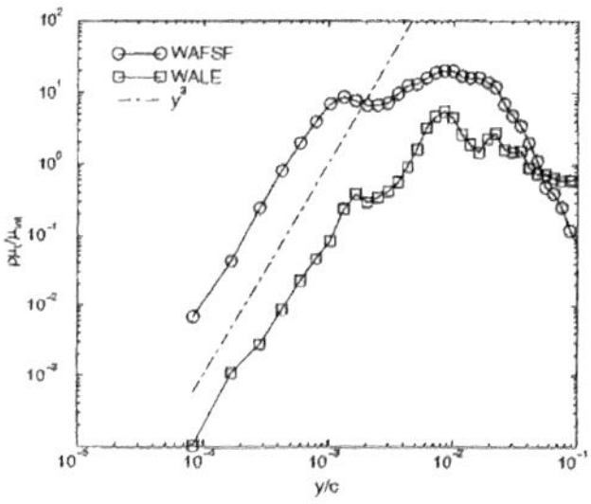

Figure 75 CERFACS: Instantaneous, spanwise-averaged eddy-viscosity profiles at $x/c = 0.7$ on the suction side for Simu 1. μ_{inf}= laminar viscosity

Conclusions. This work produced important knowledge on both the simulation and the modelling aspects. As far as pure results are concerned, we have obtained a database of simulations using both RANS and LES approaches on both sharp and blunt trailing edges. The average solutions on the sharp trailing edge exhibit no separation at all whereas a separation is seen (albeit somewhat too large) for the blunt trailing edge, but this difference may also be partly due to the different grid and numerical procedure used. The estimation of the fluctuating velocity is very sensitive to the resolution and follows a known trend: coarse resolution leads to fluctuations of excessive intensity, even with the use of wall functions.

From a purely numerical point of view the project has prompted the development of the wiggle detector procedure that seems useful in removing oscillations near "energetic events" without the use of upwind biased schemes or the use of any systematic artificial viscosity. The use of wall functions is also definitively a good way to reduce the CPU time whereas the use of wall resolved meshes leads to simulations which are too expensive. This observation suggests the promotion of new studies combining LES and wall functions or alternative methods involving RANS and LES coupling.

4.5 Contribution by Dassault-Aviation

Author: M. Ravachol

Description of the Navier-Stokes code. The Navier Stokes code used for this study is an LES extension of the unstructured finite-element compressible Navier-Stokes industrial code from Dassault-Aviation. Dassault Aviation's Navier-Stokes code uses a finite element approach, based on a symmetric form of the equations written in terms of entropy variables. The advantages of this change of variables are numerous: in addition to the strong mathematical and numerical coherence they provide

- dimensionally correct dot product,
- symmetric operators with positivity properties,
- efficient preconditioning.

Entropy variables yield further improvements over the usual conservation variables, in particular in the context of chemically reacting flows. A more detailed description of the global code is given in Chalot *et al.* [64]. For a more extensive description of the use of entropy variables, see Hughes *et al.* [65].

The symmetric Navier-Stokes equation. As a starting point, we consider the Euler and Navier-Stokes equations written in conservative form:

$$U_{,t} + F^{adv}_{i,i} = F^{diff}_{i,i}$$

where U is the vector of conservative variables; F^{adv}_i and F^{diff}_i are, respectively, the advective and the diffusive fluxes in the ith-direction. Subscript commas denote partial differentiation and repeated indices indicate summation.

The above equation can be rewritten in quasi-linear form:

$$U_{,t} + A_i U_{,i} = (K_{ij} U_{,j})_{,i}$$

where $bA_i = F^{adv}_{i,U}$ is the ith advective Jacobian matrix, and $K = [K_{ij}]$ is the diffusivity matrix, defined by $F^{diff}_i = K_{ij} U_{,j}$. The A_i's and K do not possess any particular property of symmetry or positiveness.

We now introduce a new set of variables,

$$V^T = \frac{\partial \mathcal{H}}{\partial U}$$

where $\mathcal{H}$ is the generalized entropy function given by

$$\mathcal{H} = \mathcal{H}(U) = -\rho s$$

and s is the thermodynamic entropy per unit mass. Under the change of variables $U \mapsto V$, the quasi-linear form of the Navier-Stokes equations becomes:

$$\tilde{A}_0 V_{,t} + \tilde{A}_i V_{,i} = (\tilde{K}_{ij} V_{,j})_{,i}$$

where

$$\begin{aligned} \tilde{A}_0 &= U_{,V}, \\ \tilde{A}_i &= A_i \tilde{A}_0, \\ \tilde{K}_{ij} &= K_{ij} \tilde{A}_0. \end{aligned}$$

The Riemannian metric tensor $\tilde{A}_0$ is symmetric positive-definite; the $\tilde{A}_i$'s are symmetric; and $\tilde{K} = [\tilde{K}_{ij}]$ is symmetric positive-semidefinite. In view of these properties, the new form of the Navier-Stokes equations is referred to as a symmetric advective-diffusive system.

Taking the dot product of the new system of equations with the vector V yields the Clausius-Duhem inequality, which constitutes the basic nonlinear stability condition for the solutions. This fundamental property is inherited by appropriately defined finite element methods, such as the one described in the next section.

The Galerkin/least-squares formulation. The Galerkin/least-squares (GLS) formulation introduced by Hughes [66] and Johnson [67], is a full space-time finite element technique employing the discontinuous Galerkin method in time. The least-squares operator ensures good stability characteristics while retaining a high level of accuracy. The local control of the solution in the vicinity of sharp gradients is further enhanced by the use of a nonlinear discontinuity-capturing operator.

We consider the time interval $I =]0, T[$, which we subdivide into N intervals $I_n =]t_n, t_{n+1}[$, $n = 0, \ldots, N-1$. Let $Q_n = \Omega \times I_n$ and $P_n = \Gamma \times I_n$ where Ω is the spatial domain of interest, and Γ is its boundary. In turn, the space-time "slab" Q_n is tiled by $(n_{\mathrm{el}})_n$ elements Q_n^e. A constant-in-time scheme is used, which means that a P0 interpolation in time is assumed for the approximate solution on slab I_n (see [68]). Consequently, the Galerkin/least-squares variational problem can be stated as :

Within each Q_n, $n = 0, \ldots, N-1$, find $V^h \in \mathcal{S}_n^h$ (trial function space), such that for all $W^h \in \mathcal{V}_n^h$ (weighting function space), the following equation holds:

$$\int_{Q_n} \left(-W^h_{,t} \cdot U(V^h) - W^h_{,i} \cdot F^{adv}_i(V^h) + W^h_{,i} \cdot \tilde{K}_{ij} V^h_{,j}\right) dQ$$

$$+ \int_{\Omega} \left(W^h(t^-_{n+1}) \cdot U\left(V^h(t^-_{n+1})\right)\right) - W^h(t^+_n) \cdot U\left(V^h(t^-_n)\right) d\Omega$$

$$+ \sum_{e=1}^{(n_{el})_n} \int_{Q^e_n} (\mathcal{L}W^h) \cdot \tau\left(\mathcal{L}V^h\right) dQ + \sum_{e=1}^{(n_{el})_n} \int_{Q^e_n} \nu^h g^{ij} W^h_{,i} \cdot \tilde{A}_0 V^h_{,j}\, dQ$$

$$= \int_{P_n} W^h \cdot \left(-F^{adv}_i(V^h) + F^{diff}_i(V^h)\right) n_i\, dP$$

The first and last integrals represent the Galerkin formulation written in integrated-by-parts form. The solution space consists of piecewise polynomials which are continuous in space, but are discontinuous across time slabs. Continuity in time is weakly enforced by the second integral which contributes to the jump condition between two contiguous slabs, with

$$Z^h(t^{\pm}_n) = \lim_{\varepsilon \to 0^{\pm}} Z^h(t_n + \varepsilon).$$

The third integral constitutes the least-squares operator where $\mathcal{L}$ is defined as

$$\mathcal{L} = \tilde{A}_0 \frac{\partial}{\partial t} + \tilde{A}_i \frac{\partial}{\partial x_i} - \frac{\partial}{\partial x_i}(\tilde{K}_{ij} \frac{\partial}{\partial x_j}).$$

τ is a symmetric matrix. The fourth integral is the nonlinear discontinuity-capturing operator, which is designed to control oscillations about discontinuities, without upsetting higher-order accuracy in smooth regions. g^{ij} is the contravariant metric tensor defined by

$$[g^{ij}] = [\xi_{,i} \cdot \xi_{,j}]^{-1}$$

where $\xi = \xi(x)$ is the inverse isoparametric element mapping, and ν^h is a scalar-valued homogeneous function of the residual $\mathcal{L}V^h$. The discontinuity capturing factor ν^h is is an extension of that introduced by Hughes, Mallet, and Franca [69]. In the present work, the whole term is left out due to the regularity of the considered problems.

A key ingredient to the formulation is its consistency. This constitutes an essential property in order to attain higher-order spatial convergence.

Solution strategy. Convergence to steady state of the compressible Navier Stokes equations is achieved through a fully-implicit iterative time-marching procedure based on the GMRES algorithm.

A low-storage extension based solely on residual evaluations was developed by Johan [70], [66]. It is particularly adapted to parallel processing, where the linear solver often constitutes a painful bottleneck. This algorithm has proven extremely efficient on many scalar or vector architectures. It is the basis on which all the LES developments we are going to describe, were built up.

Turbulence Models and Implementation. Different turbulence models are available, including the one-equation Spalart-Allmaras model [71], several versions of k-ε models with wall functions or low Reynolds number correction (see Chen and Patel [72] and C. Kasbarian [73]). The solution of the equations of the mean flow and the additional equations of turbulence is uncoupled, mainly in order to keep a "reasonable" size for the linear system to solve. At each time step, the equations of the mean flow are solved first, and then the additional turbulent equations, which allows a great flexibility in the turbulence models. A robust and accurate technique was developed to enforce the positivity of the turbulence variables, for which a P1 element is used.

Towards LES. The aim of LES is to reduce the number of grid points required by DNS calculations of turbulent flows. The mesh size can be further reduced by the use of unstructuredness.

The code described in the preceding section derives from its finite element structure the capacity of treating any unstructured combination of tetrahedra and hexahedra. Any order of accuracy could be achieved with appropriate elements; in practice, linear elements are used which yield global second-order accuracy in space. A detailed description of LES implementation and LES sub-grid scale models available can be found in [74]. This kind of accuracy proves to suffice for steady state calculations, but must still be critically evaluated in an LES framework.

Explicit versus implicit. In order to assess the space accuracy, any problem which may arise from a lack of time accuracy must be removed. Thus, we replaced the first-order constant-in-time scheme described above with a classical fourth-order Runge-Kutta scheme (besides making the time-step global). In addition to being more accurate, this explicit approach alleviates the need of solving any linear system. In fact, no linear system solution can be guaranteed free of either numerical truncation error or incomplete convergence side effect. The use of a higher-order explicit scheme takes away all these potential problems at once, and enables the trouble-free evaluation of the Galerkin/least-squares operator.

On the time-consistency of the Galerkin/least-squares operator. The Galerkin/least-squares operator reads:

$$\sum_{e=1}^{(n_{el})_n} \int_{Q_n^e} \left(\mathcal{L}W^h\right) \cdot \tau \left(\mathcal{L}V^h\right) \, dQ$$

This term is residual-based. This consistency property enforces higher-order accuracy in both space and time. For steady application, the term $\tilde{A}_0 \partial/\partial t$ is dropped out from the Navier-Stokes operator $\mathcal{L}$. In order to retain time accuracy, it is crucial that this term be reinstated. In the framework of a Runge-Kutta scheme, we stress the fact that this time derivative must only be computed with final step solutions, and not with indermediate predictors; it is computed at the first Runge-Kutta step with the current solution and kept constant for all the remaining Runge-Kutta steps.

A modification of the intrinsic time-scale matrix τ brings also additional time ingredients in the construction of the operator: the effect of the Galerkin/least-squares term is reduced with the time-step.

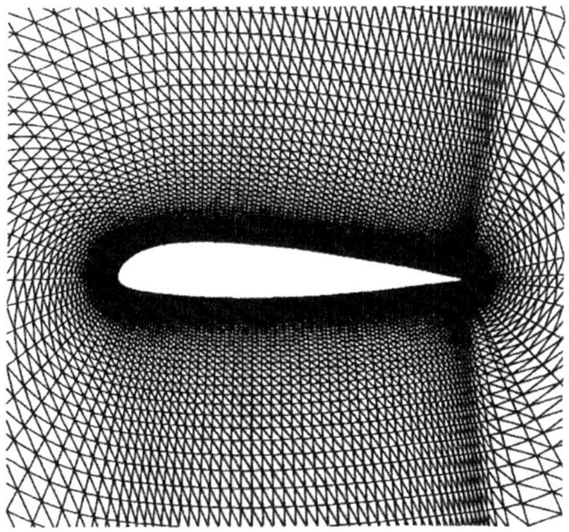

Figure 76 Global mesh for each plane

Numerical integration rule. All the integrals are numerically evaluated using Gaussian quadrature rules. As mentioned earlier, the time-dependent term and the Euler flux are integrated by parts, which preserves the conservation under reduced integration. However, although out-of-the-box finite element theory would recommend interpolating the solution variables, viz. the entropy variables V, at the Gaussian integration points, it turns out that the interpolation of primitive variables $Q^T = \{\rho, u, p\}$ gives better results. This is especially true for phenomona driven by minute pressure variations.

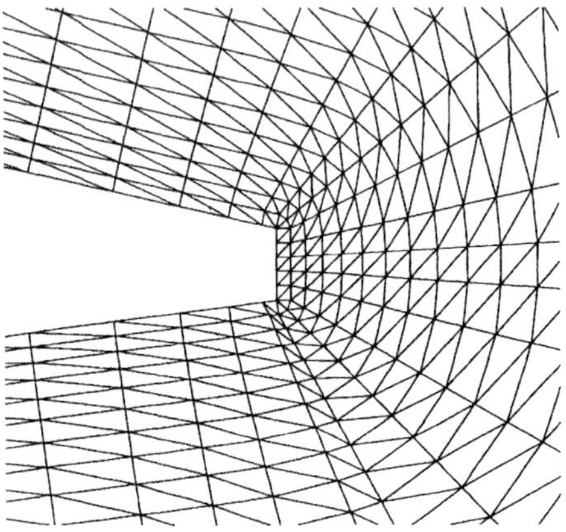

Figure 77 Zoom around the trailing edge

Application to the A-airfoil. Three different computations have been performed on this case: a RANS computation using the $k - \varepsilon$ model on the 2D geometry and two LES computations with a change in the subgrid-scale model (Smagorinsky and selective Smagorinsky). All the computations have been performed with wall functions with the same grid resolution normal to the wall (first mesh point off the wall around $y^+ = 100$). No particular measure was undertaken for the transition from laminar to turbulent flow, especially because of the use of wall functions. The near-wall resolution would not have been sufficient for capturing the transition bubble observed during the expriment. Furthermore, as currently available procedures involve additional parameters, taking transition into account would have required additional computations. Hence, in both LES and RANS the boundary layer is probably fully turbulent right from the leading edge.

The mesh. For this computation we used an ***unstructured grid*** around the airfoil. Starting from a 2D mesh (used for the RANS calculation), the 3D grid was constructured by reproducing 25 planes of the 2D mesh in the spanwise direction; the spanwise width being 3% of the chord length. The main characteristics of the grids are :

- 8409 mesh points for each plane (i.e. 210 225 points in the global mesh)
- The ***actual*** geometry is meshed : ***blunt*** trail edge

A global view of the mesh in a plane is presented in Fig. 76 and a zoom around the trailing edge is shown in Fig. 77.

In Fig. 77, one can notice that 5 points are used to describe the trailing edge of the airfoil.

Initialization of the LES computation. The computation was initialized using a snapshot from an LES result provided by CERFACS (C. Weber computation [75]). This computation was carried out on a ***structured grid*** and on a different mesh. Thus the solution needed to be interpolated on the mesh used by Dassault. No noise was added during the computation or at the initialization stage.

Computation and sampling. For this computation we used the SUPG/Least-squares stabilization which is the base of the solution method developed at Dassault-Aviation for the Navier-Stokes equations (see [64]). It is worth mentioning that this method is a stabilized method, thus not a centered method. Whilst this method inherentently adds some artificial dissipation, it is not a TVD method and therefore the numerical dissipation cannot be viewed as a SGS model. For LES we need to add a SGS model to the SUPG stabilization. The present computation was performed using a law of the wall and the standard as well as the selective Smagorinsky sub-grid scale model.

The version of the law of the wall is an instantaneous Log law which is imposed at each time step:

$$u^+ = \frac{1}{\chi} Log(y^+) + B$$

This boundary condition is implemented as follows: the velocity is prescribed at the nodes lying on the wall by inverting the law of the wall for the first point just off the wall.

The selective Smagorinsky model used for the computation is easily derived from the Smagorinsky model and can be summarized as follows:

$$\nu_t^{\mathrm{sel}} = F \nu_t^{\mathrm{Smago}}$$

where ν_t^{Smago} is the standard Smagorinsky viscosity and F the selective function. This function turns out to be a "three-dimensionnality" sensor for the flow and is constructed as follows:

$$F = \begin{cases} 0 \text{ if } \alpha \leq 20^\circ \\ 1 \text{ otherwise} \end{cases}$$

where α denotes the angle between the vorticity vector at the considered element e_i and the average of the vorticity vectors at the elements around e_i. See [74] for more details about the sub-grid scale models used.

Each LES computation (Smagorinsky and Selective Smagorinsky) was carried out for 10 time units ($\frac{c}{U_\infty}$) starting from the *same initial condition.* In both cases, the first 4 time units were used to develop the flow and the averaging was carried out over the last 6 time units.

Comparison of LES results using different SGS models. We compare here the results of the LES simulations obtained with the Smagorinsky model and with its selective version. All the solid curves refer to the Smagorinksy model and the dashied curves to the selective Smagorinky model.

Pressure coefficient. In Fig. 78 we present the average pressure coefficient from the two LES simulations. The two main differences are at mid-chord location ($0.4 \leq x/c \leq 0.6$) and at the trailing edge. At the mid-chord, the Smagorinksy model generates some oscillations in the pressure distribution, the selective model is in better agreement with the experiment but still gives non-physical results. A possible explanation is that the creation of eddy viscosity is too strong in this region, even if it is damped in the selective case. It seems that the LES generates a kind of numerical transition around $x/c = 0.3$. In the experiment, there is a laminar bubble on the suction side of the airfoil and the transition occurs in the bubble. This feature is not captured by the LES due to the lack of near wall resolution of the mesh used. Moreover, it seems difficult for the wall function to model such an elaborated transition process. For the section between $x/c = 0.6$ and $x/c = 0.8$ the two simulations give similar results and the results are again different at the trailing edge: the selective model generates oscillations and the standard Smagorinsky model agrees better with the experimental data.

Velocity profiles. In Fig. 79 we present a comparison between the LES results and the experimental results for the sections $x/c = 0.3$, $x/c = 0.775$, $x/c = 0.93$ and $x/ = 0.99$. The first two profiles correspond to sections before the separation located in the experiment around $x/c = 0.85$. For the first profile: $x/c = 0.3$ the two models already give different results; the boundary layer obtained with the Smagorinsky model seems to be thicker. For the profile at $x/c = 0.775$, the difference is larger: the selective model gives results close to the experimental data and the Smagorinsky model departs from them. This trend is confirmed by comparisons at others locations that are not presented here.

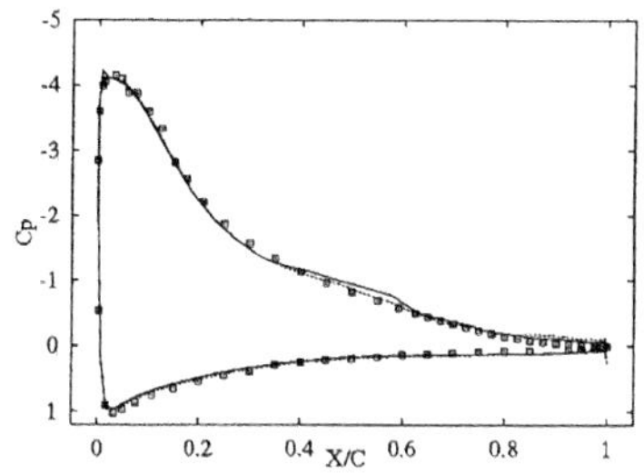

Figure 78 Dassault. Pressure Coefficient: ——— Smagorinsky, ... selective Smagorinsky, □ experiment

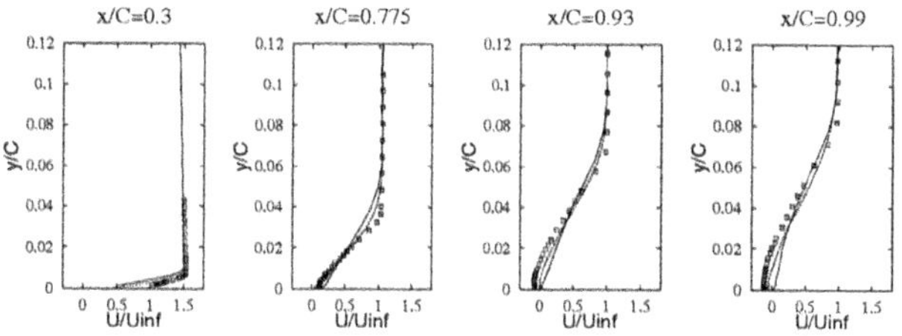

Figure 79 Dassault. LES velocity profile: ——— Smagorinsky, ... selective Smagorinsky, □ experiment

The two last profiles at $x/c = 0.93$ and $x/c = 0.99$ in Fig. 79 compare LES results and the experiment in the separated region. For these profiles, the results obtained with the selective model are in better agreement with the experiment than the results obtained with the Smagorinsky model. The Smagorinsky model does not show any back flow.

This is confirmed by the zoom of the velocity profiles (see Fig. 80). The selective model is in good agreement with the experimental data even for the back-flow region but the Smagorinksy model does not predict any back flow at all.

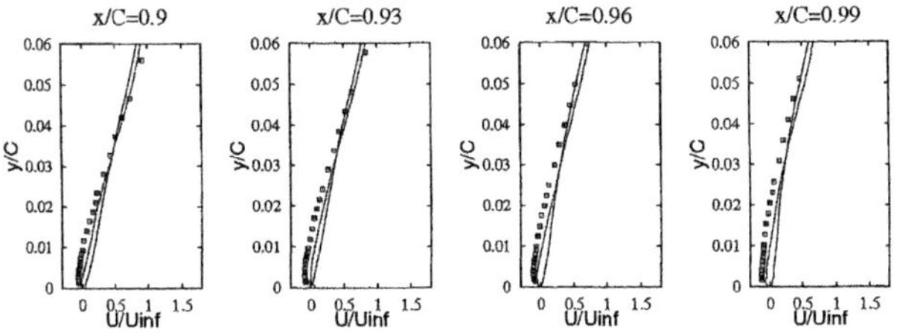

Figure 80 Dassault. Zoom close to the wall near the trailing edge of the streamwise velocity profiles: —— Smagorinsky, ... selective Smagorinsky, □ experiment

Profiles of Fluctuations. In this section, we present a comparison between three components of the Reynolds stress tensor obtained with the Smagorinsky model computation and the experimental results for the sections x/c=0.3, 0.7, and 0.93. The variables presented are: $\langle uu\rangle/U_\infty^2$, $\langle vv\rangle/U_\infty^2$ and $\langle uv\rangle/U_\infty^2$. These three components are estimated by averaging the resolved velocity fields provided by the LES calculation (thus the contributions of the modelled subgrid components are not taken into account). Due to a computational problem (disk storage not sufficient), data are not available for the selective Smagorinsky LES computation, which is unfortunate since it was shown that with this model the results for the first-order moments are in better agreement with the experimental data.

Before going into details for each section, we must first warn that the number of samples used for time-averaging may not be sufficient to have an accurate estimation of the statistical values of the fluctuating quantities, as the averaging time was calibrated for the main velocity profiles.

At the first profile x/c=0.3 (Fig. 81), the turbulent boundary layer is not well developed yet (the transition point was not specified in the computation).

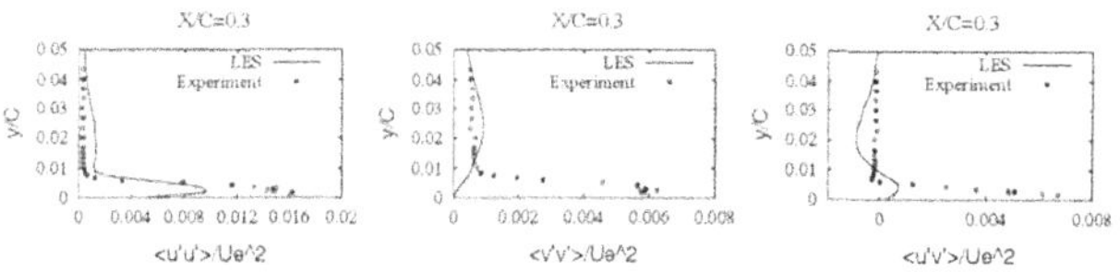

Figure 81 Dassault. Reynolds stresses $\langle uu \rangle / U_\infty^2$, $\langle vv \rangle / U_\infty^2$, $\langle uv \rangle / U_\infty^2$ at $x/c = 0.3$. Smagorinsky model.

As a result, the maximum value of the three Reynolds stress components is very close to the wall. The comparison is quite good for $\langle uu \rangle / U_\infty^2$, even if the sharp peak near the wall is underestimated. The numerical results may be improved by taking the subgrid contribution into account. For the other components, the comparison is not so good. This may be linked to the the fact that the viscous sublayer is not resolved but wall functions are used.

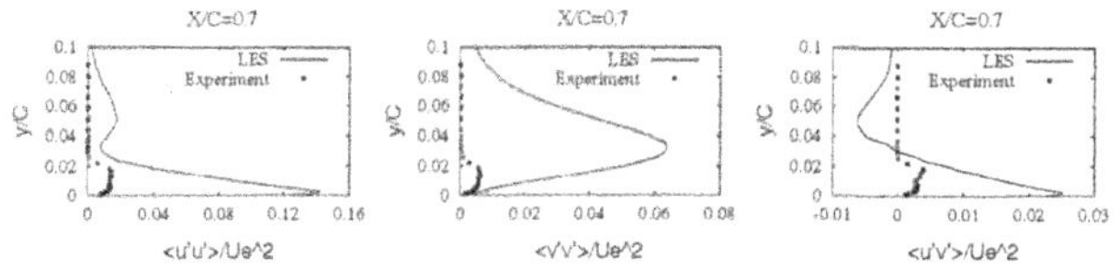

Figure 82 Dassault. Reynolds stresses $\langle uu \rangle / U_\infty^2$, $\langle vv \rangle / U_\infty^2$, $\langle uv \rangle / U_\infty^2$ at $x/c = 0.7$. Smagorinsky model.

At x/c=0.7 (Fig. 82), the profiles of $\langle uu \rangle / U_\infty^2$, $\langle vv \rangle / U_\infty^2$ and $\langle uv \rangle / U_\infty^2$ are quite curious: the experimental data indicate that the Reynolds stress components are fairly small, whereas the LES results show a significant contribution. Too coarse a mesh may be the reason for these differences. This may indicate that from $y/c \approx 0.03$ on, the mesh resolution is not sufficient to accurately compute the fluctuations.

The same remark applies to the $\langle uu \rangle / U_\infty^2$ profiles at x/c=0.93 (Fig. 83). The second peak is non physical, and also may result from an insufficient resolution in this area. For the shear-stress component, the comparison is better, even if the maximum is still overestimated.

Evaluation of the SGS models. From the results presented above, it is clear that the selective version of the Smagorinsky model performs better than the

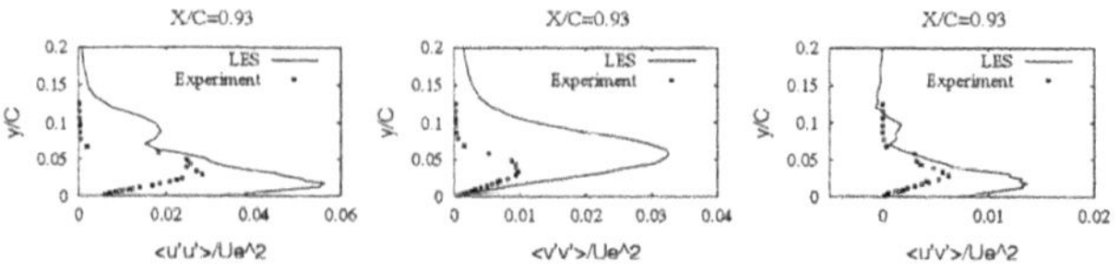

Figure 83 Dassault. Reynolds stresses $\langle uu \rangle / U_\infty^2$, $\langle vv \rangle / U_\infty^2$, $\langle uv \rangle / U_\infty^2$ at $x/c = 0.93$. Smagorinsky model.

original Smagorinsky model. The latter seems to generate too high a viscosity level preventing the flow to separate in the trailing edge region. The selection function inhibits the creation of eddy viscosity in the region where the flow is two-dimensional and thus the selective version of the Smagorinsky model gives an overall better agreement with the experiment.

Moreover, the extra CPU cost needed to evaluate the selection function, although not being negligible since it necessitates to average around each element, is still acceptable in view of the improvement of the results.

Comparison between RANS and LES. In this section we compare the results obtained from a 2D RANS calculation performed with the $k-\varepsilon$ model with wall function and the 'best' LES we have obtained (i.e. with the selective Smagorinsky model). The RANS computation is obtained for the same conditions (Mach number, Reynolds number and angle of attack) for which the steady state is reached. Due to the use of wall functions, the flow is always turbulent and no particular measure was undertaken to enforce transition (as for the LES computation).

The 2D mesh used for the RANS calculation is built by extracting a section of the mesh used for the LES, thus the two computations are performed with the same resolution normal to the airfoil.

The comparison deals only with mean-average values, as profiles of fluctuation quantities are not available for the selective Smagorinsky LES computation (due to the computational problem mentioned above). As will be shown, the mean quantities (pressure and velocity) are better predicted by the LES computation than by the RANS computation, so that a comparison of first-order moments may be sufficient to highlight the improvements due to LES.

Pressure coefficient. In Fig. 84 the LES computations gives an overall better agreement with the experimental results than the RANS computation. In particular, the suction peak at the leading edge of the airfoil is better predicted by the LES whilst RANS overpredicts the peak. Moreover, it is interesting to look at the pressure side of the airfoil. The LES computation compares extremely well with the experimental data in comparison with the RANS solution which shows some discrepancies. This indicates that, even though the specified angle of attack is the same for RANS and LES, the *actual* angle of attack is different for the two computations; in other words, the circulation around the airfoil is different and the LES seems to agree better with the experiment.

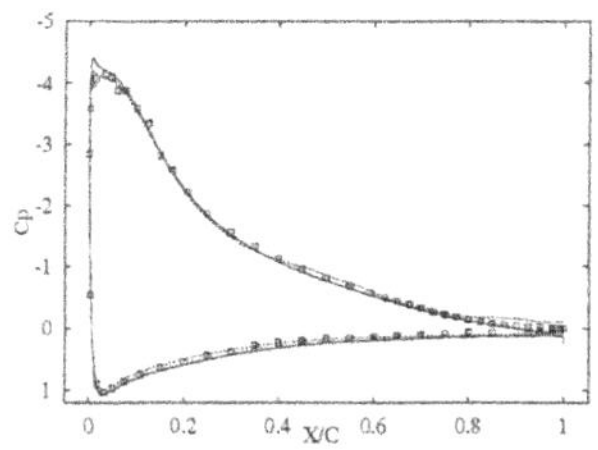

Figure 84 Dassault. Pressure Coefficient: —— RANS $k-\varepsilon$, ... selective Smagorinsky, □ experiment

If we concentrate on the LES results, they are far from being perfect. For instance, the C_p profile between stations $x/c = 0.4$ and $x/c = 0.5$ is not physical and the pressure plateau for $x/c \geq 0.8$ is overpredicted. The first deficiency may be due to the selective function, $F = 0$ or 1. As mentioned above, the selective function damps the creation of eddy viscosity in the non-turbulent region, but the combined effect of the brutal switch and the coarse mesh in the near-wall region may lead to a "numerical transition" which can lead to a discontinuity in the wall pressure distribution. The overprediction of the pressure plateau may also be explained by "numerical transition". In the experiment the circulation around the airfoil (and thus the trailing edge pressure) depends on the laminar bubble. Since the LES does find a separation at the trailing edge but no laminar bubble, the circulation is different and balanced by a different distribution at the trailing edge.

Velocity field. In Fig. 86 we compare contours of the u component of the velocity field for the RANS solution and the averaged LES solution. The overall agreement between the two solutions is pretty good. The main differences are in the trailing edge region where the LES solution produces a thicker boundary layer and in the inviscid part of the flow where the LES solution is more "noisy". We want to recall that no noise was added during the computation and the "wiggles" observed are a by-product of the LES simulation.

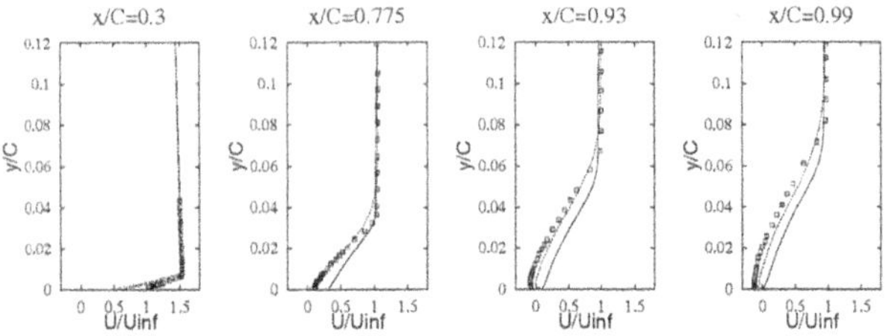

Figure 85 Dassault. Profiles of streamwise velocity: —— RANS $k-\varepsilon$, ... selective Smagorinsky, □ experiment

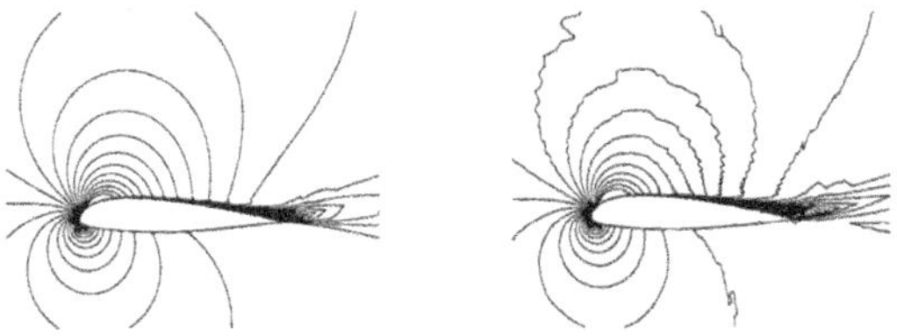

Figure 86 Dassault. Contours of u-velocity: RANS (left) and Averaged LES (right)

Velocity profiles. In this section we compare the mean velocity profiles obtained from the RANS computation and the LES computation against the experimental data. In Fig. 85 the profiles for stations $x/c = 0.3$, $x/c = 0.775$, $x/c = 0.93$ and $x/c = 0.99$ are presented. The two first profiles deal with sections located before experimental separation. Except for station $x/c = 0.4$ (not presented here), the LES solution agrees better with the experimental results than the RANS computation. In particular, the outer part of the boundary layer is better described in the LES solution. At station $x/c = 0.4$, as explained earlier for the pressure coefficient the discrepancy observed may be linked to a deficiency in the SGS model.

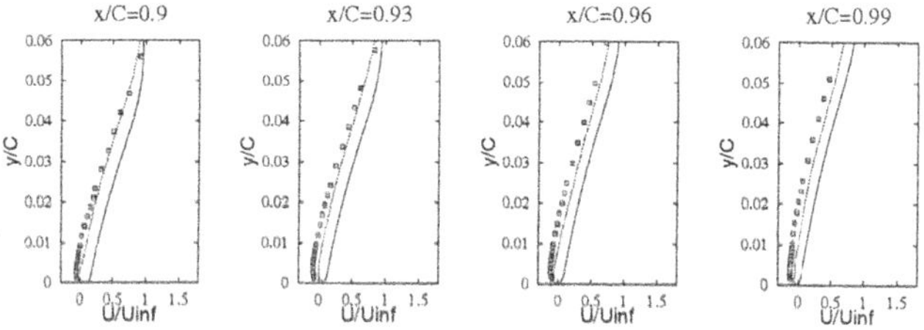

Figure 87 Dassault. Profiles of stream velocity, zoom near trailing edge: ——— RANS $k-\epsilon$, ... selective Smagorinsky, □ experiment

The two last sections of the profiles of Fig. 85 are in the trailing edge region (and a zoom in Fig. 87). The LES computation agrees very well with the experiment and the separation at the trailing edge is captured. In the separated region, the agreement between the LES result and the experimental data is good. If one keeps in mind that the velocity boundary condition is set using a log law which is usually considered not to work for separated flows, the agreement between the LES and the experiment is almost surprisingly good in the back-flow region. Maybe the usual observation concerning the law of the wall derived from the RANS experience cannot be extended so easily to LES.

Conclusions. These LES simulations performed using the industrial code from Dassault-Aviation assess the feasibility of LES for the flow around an airfoil configuration involving complex physical behavior at a "reasonnable" cost. The main features of the flow are correctly described: pressure distribution and velocity profiles. Even if the results are far from being perfect, they are encouraging in the sense that most of the steady RANS computations cannot predict the separation at the trailing edge for this case; the LES does predict a separation and the velocity profiles compare well with the experiment in the separated region. The differences observed concern the physics that could not predicted with the approach used here: laminar bubble at the leading edge with transition in the bubble.

Nevertheless, this computation performed using an "industrial" approach with minimal mesh and wall function shows the potential of LES in terms of its ability to describe separated flows.

4.6 Contribution by FLUENT

Authors: F. Liu and D. Cokljat

Introduction. This contribution summarizes the results obtained using the detached-eddy simulation (DES hereafter) technique for the prediction of the flow around the high-lift A-airfoil at high Reynolds number. DES has been adopted here as a more practical alternative to the LES simulations of high-Re number flows around airfoil-type geometries. The LES results obtained during the duration of the LESFOIL project have indicated that in order to obtain a good accuracy it is necessary to have a high resolution grid. In that respect, the only successful simulation (to our knowledge) of the currently considered A-airfoil flow with an angle of attack of $\alpha = 13.3°$ required 7.2 million cells (see [?] of which an abbreviated version is given in Section 4.8). This is despite the fact that the authors considered a large part of domain to be two dimensional. Furthermore, the domain in the spanwise direction had an extent of only 1.2 % of the chord length. Thus a successful LES simulation of a high Reynolds number flow around the airfoil will require a mesh of substantial size. Unfortunately, such a conclusion consequently puts a question mark on the practicability of LES for the type of flows considered in the present studies for every-day industrial usage.

From that point of view, DES can be considered as a more practical alternative to the LES simulation of flows around high-Re high-lift airfoils and may offer a hope towards capturing the physics of this highly challenging flow in an affordable manner. DES, adopted in this study, employs a modified version of the standard one-equation Spalart-Allmaras turbulence model [55].

Model Description. The Spalart-Allmaras (SA) model solves a transport equation for the turbulent kinematic viscosity, $\tilde{\nu}$:

$$\rho\frac{D\tilde{\nu}}{Dt} = \frac{\partial}{\partial x_j}\left(\frac{\mu + \rho\tilde{\nu}}{\sigma_{\tilde{\nu}}}\frac{\partial\tilde{\nu}}{\partial x_j}\right) + \frac{C_{b2}\rho}{\sigma_{\tilde{\nu}}}\left(\frac{\partial\tilde{\nu}}{\partial x_j}\right)^2 + G_\nu - Y_\nu \tag{14}$$

where G_ν and Y_ν are the production and destruction term respectively. The turbulent viscosity, μ_t, is computed from

$$\mu_t = \rho\,\tilde{\nu}\,f_1 \tag{15}$$

where f_1 is the viscous damping function. The production term is modeled as:

$$G_\nu = C_{b1}\,\rho\left(S + \frac{\tilde{\nu}}{\kappa^2 d^2}f_2\right)\tilde{\nu} \tag{16}$$

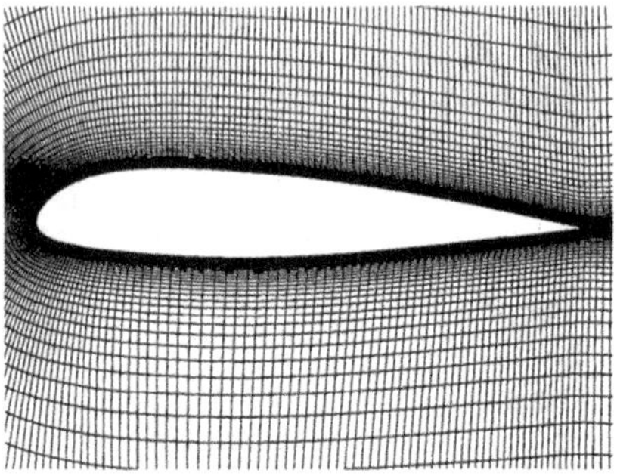

Figure 88 FLUENT: Grid distribution in $x - y$ plane.

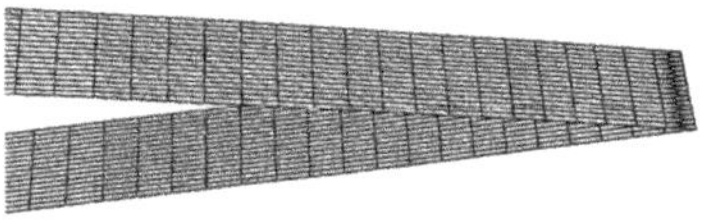

Figure 89 FLUENT: Grid resolution in span-wise direction.

where S is a scalar measure of the deformation tensor, d a length scale and f_2 the production term damping function. The destruction term is modeled as:

$$Y_\nu = C_{w1} \rho f_w \left(\frac{\tilde{\nu}}{d} \right)^2 \qquad (17)$$

with f_w defined as destruction term damping function. The definition for all damping functions shown above is the same as in standard SA model [55], hence, they are not repeated here. The standard SA model [55] uses the distance to the closest wall as a definition for the length scale d and it is obvious from the above equations that it plays a major role in determining the level of production and destruction of turbulent viscosity. The DES model, proposed in [24], replaces d everywhere in the above equations with a new length scale, $\tilde{d}$, defined as:

$$\tilde{d} = min(\, d,\, C_{des}\, \Delta) \qquad (18)$$

where the grid spacing, Δ, is based on the largest grid space in either of the x, y, or z directions forming the computational cell. The empirical constant C_{des} is taken as 0.65.

In reality this formulation creates two separate regions inside the flow domain: one LES based and the other one close to the wall where the modeling is dominated by the RANS based approach. The LES region is normally associated with the high Reynolds number turbulent core region where large turbulence scales play a dominant role. In this region, the DES formulation recovers pure LES based on the one-equation sub-grid model. Close to the wall where viscous effects prevail the standard RANS formulation is recovered.

The Mesh. The geometry of the airfoil was supplied by ONERA and the computational domain stretches 10 chord lengths in the x and y directions away from the airfoil. Initially, a 2D-mesh of 360 × 64 (Fig. 88) cells was used. This mesh has the same resolution in the $x - y$ plane as the common wall-function mesh supplied for the LESFOIL project. The grid is stretched in the direction normal to the airfoil giving a finer resolution close to the wall. For 3D simulations the geometry has been extended in the span-wise direction so that it is 3% of the chord length. In the span-wise direction a uniform grid of 16 cells has been used (Fig. 89), resulting in a total grid of 368640 cells. This mesh is only half size of the common wall function mesh previously used by all partners. By reducing the span-wise extent from 12% of the chord length (as originally proposed by the partners) to 3% we can improve cell resolution even with just 16 cells (the original proposition was for 32 cells with a spanwise extent of 12%). It is worth noting that other partners have followed a similar strategy, namely Dassault used 3% of the

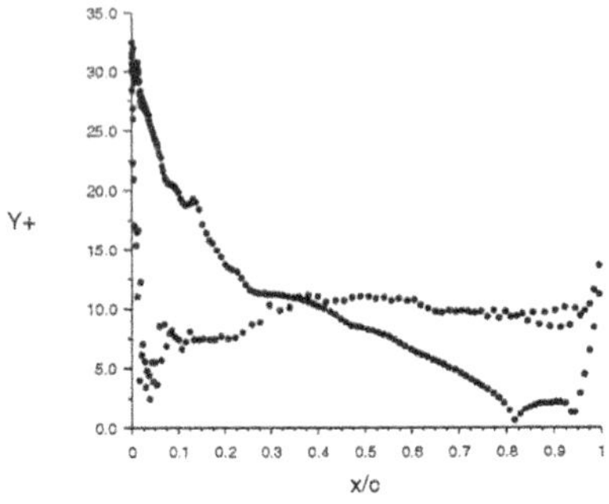

Figure 90 FLUENT: Instantaneous values of y_1^+.

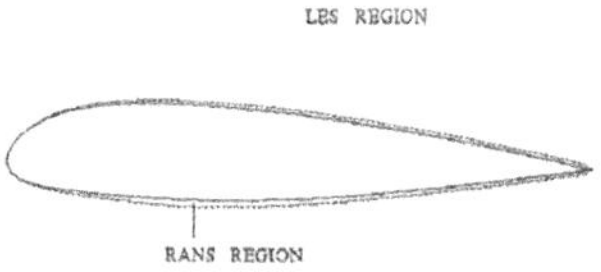

Figure 91 FLUENT: Demarcation line between RANS and LES regions.

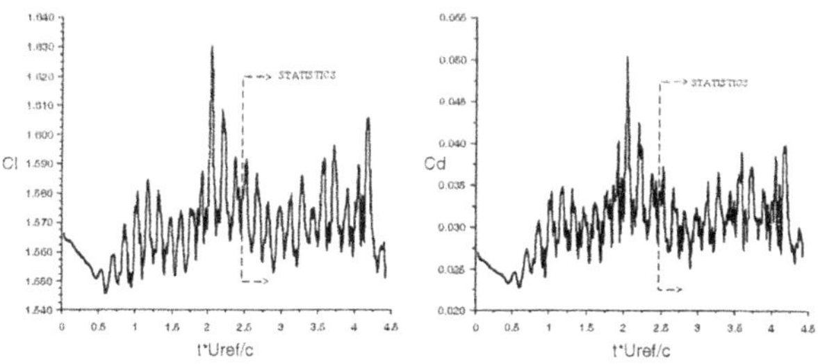

Figure 92 FLUENT: Time history of lift coefficient (**left**) and drag coefficient (**right**).

chord length in the span-wise direction while ONERA in Section 4.8 used 1.2%. It is important to note that all results presented in this report are obtained from the 3D simulation unless indicated otherwise.

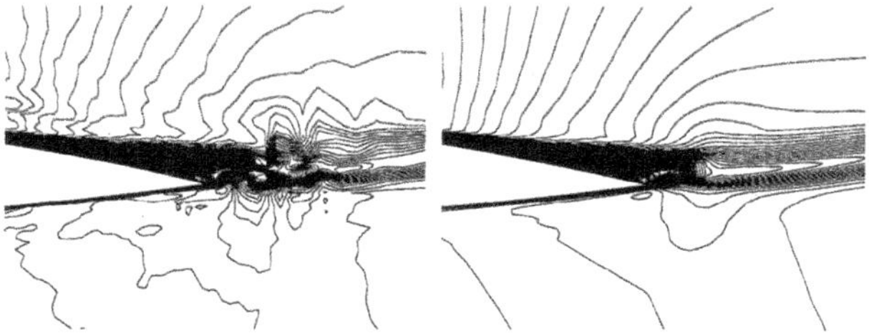

Figure 93 FLUENT: Iso-contours of streamwise velocity. **Left:** instantaneous velocity, **right:** average velocity.

Numerical Details. All calculations have been obtained using a development version of the FLUENT V5 general-purpose Control-Volume code with DES modifications. The DES runs used a second-order implicit scheme for temporal discretization and central differencing for spatial discretization. The SIMPLEC algorithm was used for the pressure-velocity coupling.

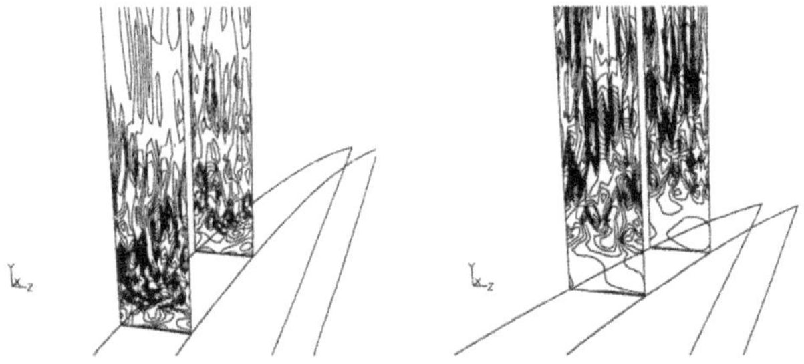

Figure 94 FLUENT: Contours of instantaneous streamwise vorticity. **Left:** $x/c = 0.3$ and 0.4, **right:** $x/c = 0.6$ and 0.7.

The velocity at the inlet is specified to its freestream values where $Ma = 0.15$ and the angle of attack is $\alpha = 13.3°$ giving $Re = 2 \times 10^6$. At the wall the common wall-function grid has typically resulted in a distance of the first grid point from the wall of $y_1^+ < 35$ as can be seen in Fig. 90 where instantaneous values of y_1^+ are displayed. Turbulence at the inlet was specified to a very low value with an intensity of 0.0001%. Time statistics have been collected over a time period of approximately $2c/U_\infty$ and the time step was set to the value of $\Delta t = 3. \times 10^{-4} c/U_\infty$ giving $CFL_{max} = 0.5$ for the present calculations. As already mentioned, the DES approach results in the creation of two distinct modeling regions (RANS and LES based) and the demarcation line between the two is displayed in Fig. 91. It is clear from this figure that the RANS region is very much confined to the near wall, whereby typically it contains $3 - 13$ cells in the vicinity of the wall. The value of y^+ at the demarcation line is different at different locations. For example, the lowest value on the suction side is $y^+ = 35$ close to the trailing edge while the highest value is $y^+ = 580$ close to the location $x/c = 0.35$.

Results. The method used to obtain results using the DES technique is the same as for LES simulations. Here we use converged steady state results obtained using the standard Spalart-Allmaras model as an initial field for the unsteady simulations. Before we start to collect statistics it is necessary that flow field is reasonably statistically stable. In order to monitor this, we plot the time history of the lift and the drag coefficient, as shown in Fig. 92. It is clear from these figures that flow instabilities start at about 0.5 of a dimensionless time unit (c/U_∞). At about $t = 2.5c/U_\infty$ the flow is already in stable quasi-periodic mode and by this time the influence of initial field is negligible on the flow behavior. From this point, we collect statistics for another $2c/U_\infty$.

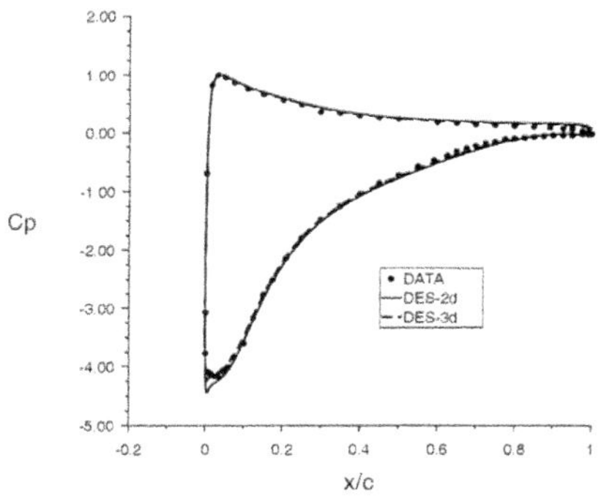

Figure 95 FLUENT: Time-averaged pressure coefficient around the airfoil.

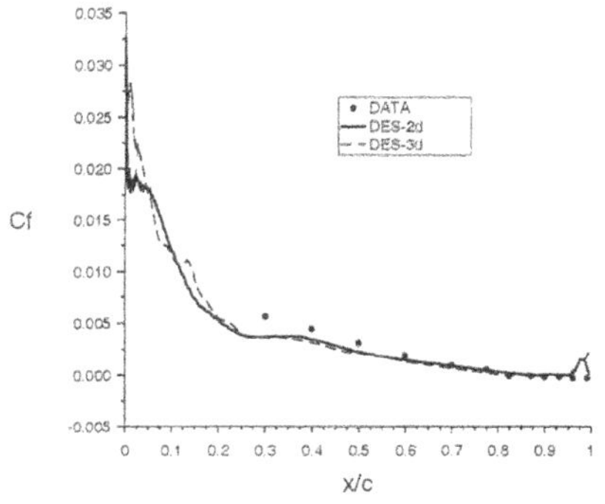

Figure 96 FLUENT: Time-averaged skin friction coefficient on the suction side of the airfoil.

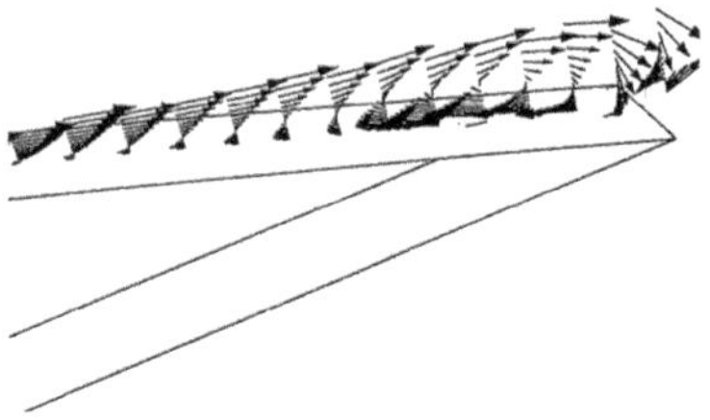

Figure 97 FLUENT: Time averaged velocity vectors at the end of the suction side.

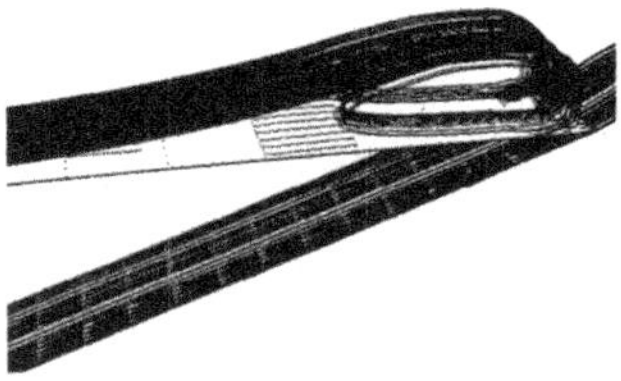

Figure 98 FLUENT: Velocity path-lines based on time-averaged velocity field.

Fig. 93 shows the instantaneous and time-averaged U-velocity contours around the trailing edge. Clearly, fluctuations are depicted in the instantaneous velocity field and it can be seen that the model captures a vortex shedding in the area immediately downstream from the trailing edge. Time-averaging leads to smooth contours and the disappearance of the vortex shedding which may be used as an indication that statistics have been collected over a reasonable time period.

Furthermore, in order to verify the 3D nature of the flow we have plotted x-vorticity contours on the suction side at four different planes, $x/c = 0.3$, 0.4, 0.6 and 0.7 in Fig. 94. The three-dimensional nature of the flow is clearly indicated. However, the small scale vortices are dominant further from the wall while closer to the wall they disappear and in particular in the region $x/c > 0.6$.

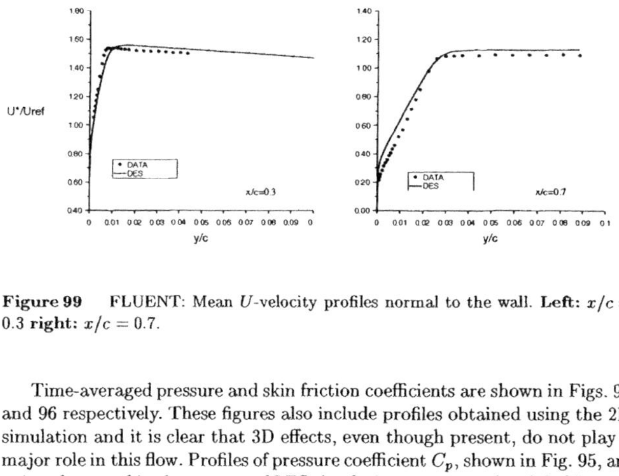

Figure 99 FLUENT: Mean U-velocity profiles normal to the wall. **Left:** $x/c = 0.3$ **right:** $x/c = 0.7$.

Time-averaged pressure and skin friction coefficients are shown in Figs. 95 and 96 respectively. These figures also include profiles obtained using the 2D simulation and it is clear that 3D effects, even though present, do not play a major role in this flow. Profiles of pressure coefficient C_p, shown in Fig. 95, are quite often used in the context of LES simulations to assess the global impact of mesh resolution and sub-grid modeling on the accuracy of the results. Therefore, it is quite often necessary to try several mesh resolutions before the C_p profile is in agreement with the data. The present results are in very good agreement with the measured value of C_p indicating that calculation of this parameter is less sensitive to the mesh resolution when coupled with RANS modeling very close to the wall. Furthermore, the skin friction results shown in Fig. 96 also show a quite satisfactory agreement with the data and in particular in the region $x/c > 0.3$. However, prior to $x/c > 0.3$ the C_f profiles do not indicate the existence of a transition region. This is not surprising given that the wall function mesh was used. Furthermore, it is important to stress that no specific treatment close to the wall was introduced in order to capture transition. Toward the end of the suction side the C_f profiles assume small negative values, indicating the existence of a recirculation zone close to the trailing edge. This can be better assessed through the velocity vectors and velocity path-lines displayed in Figs. 97 and 98 respectively. These two figures show clear recirculation on the suction side close to the trailing edge. The present recirculation covers approximately 10% of the airfoil towards the trailing edge and compared with the experimental value of about 15% it is slightly underestimated.

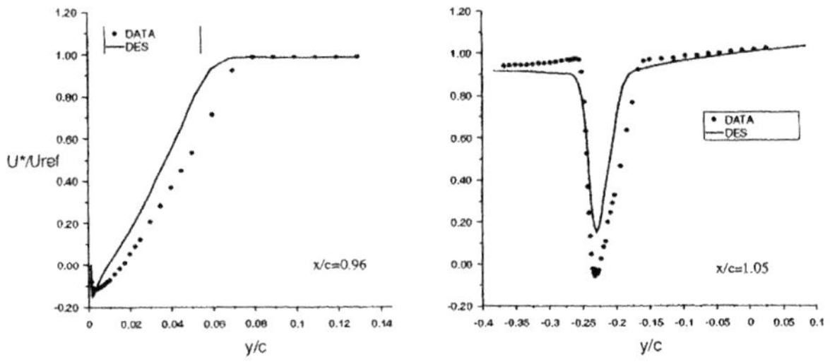

Figure 100 FLUENT: Mean U-velocity profiles normal to the wall.**Left:** $x/c = 0.96$ **right:** $x/c = 1.05$.

The predicted velocity profiles agree reasonably well with the measured data, as can be seen in Figs. 99, 100, 101 and 102. These figures display profiles at three wall locations ($x/c = 0.3$, 0.7 and 0.96) and one at $x/c = 1.05$ which is placed in the wake region immediately downstream from the trailing edge. A negative velocity region is clearly shown at x/c=0.96 by both prediction and experimental results. The discrepancies between the predicted and measured results are probably caused by deficiencies in the underlying RANS modeling used within the boundary layer. In particular a previous study [76] using the steady-state Spalart-Allmaras model has indicated similar behavior. However, the discrepancy noticed in the wall profiles does not re-occur at $x/c = 1.05$ where predictions are not influenced by any potential error that may be associated with RANS type modeling (note that this position is outside the RANS region). An excellent agreement between the predicted and measured mean velocity profiles at $x/c = 1.05$ is also accompanied with very good agreement for the RMS velocity values, as can be seen in Fig. 103. These fluctuations are the resolved ones only. Our interpretation is the following. In the wake region the resolved fluctuations constitute the main part of the total fluctuations. Close to the wall, on the other hand, the unresolved fluctuations are more important than the resolved ones which needs to be accounted for when comparing the latter to the total fluctuations measured in the experiment.

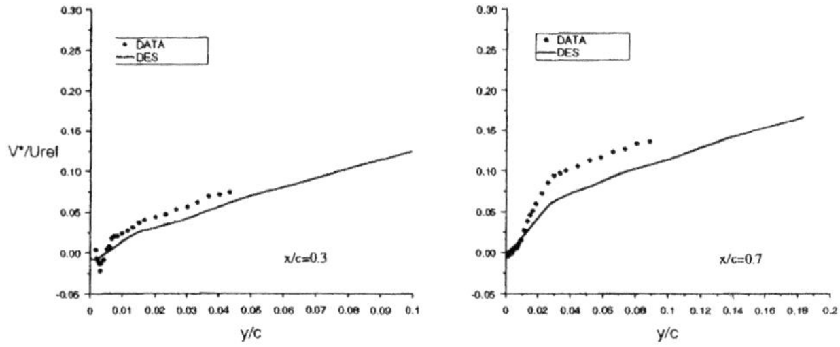

Figure 101 FLUENT: Mean V-velocity profiles normal to the wall. **Left:** $x/c = 0.3$ **right:** $x/c = 0.7$.

However, despite reasonable agreement between the data and the above shown predictions, it is important to say that the mesh used is relatively coarse and that mesh refinement will probably further improve the accuracy of the results (note that the Spalart Allmaras model is essentially a low-Re model which gives the best performance for meshes in which the distance of the wall-adjacent grid point y_1^+ is about 1). This is particularly true since according to the measured velocity fluctuations turbulence exists over the larger part of the chord length of the airfoil surface and also fairly close to the wall. This might indicate a need for better resolution closer to the wall in order to be able to correctly predict the onset position of turbulence on the suction side of the airfoil. However despite the differences, both lift and drag forces are very close to the experimental data, as shown in Table 26. It is interesting to note from this table that the 3D effects of turbulence, despite being small, still do affect the accuracy of the global lift and drag coefficients, resulting in better agreement with the data for the 3D simulations.

Table 26 FLUENT: Drag and lift coefficients for the A-airfoil.

Method	Angle of attack $\alpha = 13.3°$	
	Cl	Cd
Experiment F2	1.515	0.0307
DES (2d)	1.618	0.0287
DES (3d)	1.568	0.0303

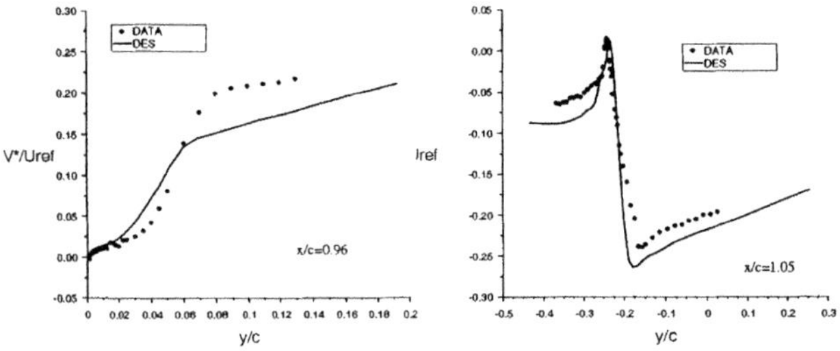

Figure 102 FLUENT: Mean V-velocity profiles normal to the wall. **Left:** $x/c = 0.96$ **right:** $x/c = 1.05$.

Conclusions. The results of Detached-Eddy Simulation of the flow around a high-lift airfoil at high Reynolds number have been presented. The common coarse mesh was used (in the $x - y$ plane) which was known from the past to lead to a very erroneous results when used in conjunction with the standard LES approach. However, this mesh when used in conjunction with the DES technique led to very encouraging results. Some of the main features of this flow, including recirculation at the end of the suction side, have been reasonably accurately predicted. This leads us to the conclusion that the DES technique may be regarded as a more practical alternative to classical LES approaches when considering the flow around high-lift airfoils at high Reynolds number.

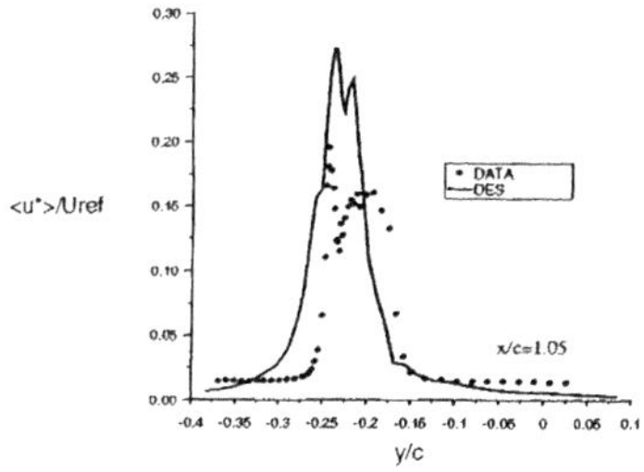

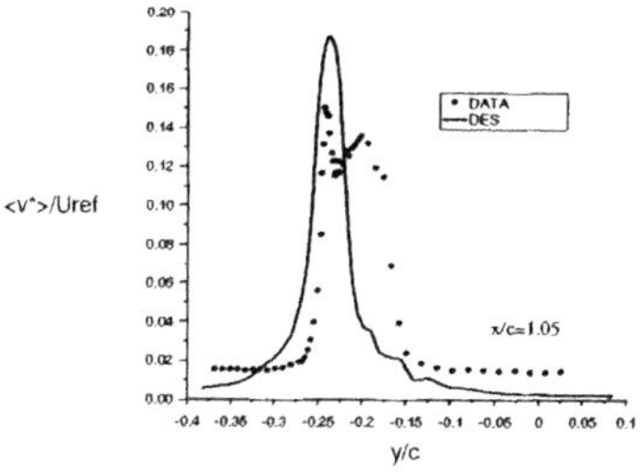

Figure 103 FLUENT: RMS velocity profiles of the resolved fluctuations at $x/c = 1.05$. **Left:** stream wise fluctuations, **right:** wall-normal fluctuations.

4.7 Contribution by University of Karlsruhe

Authors: C. Mellen, J. Fröhlich, W. Rodi

This contribution summaries the Karlsruhe studies in Task 5 over the duration of the LESFOIL project. Emphasis was placed on the subgrid-scale, near-wall and transition modelling. Furthermore, a number of additional issues have been just as influential on the direction of the Karlsruhe Task 5 effort. These revolve primarily around the need to achieve airfoil meshes with resolution sufficient to support satisfactory LES results and the challenge then inherent in computing on these meshes while avoiding prohibitive computational costs. These issues are discussed further in the sections which follow.

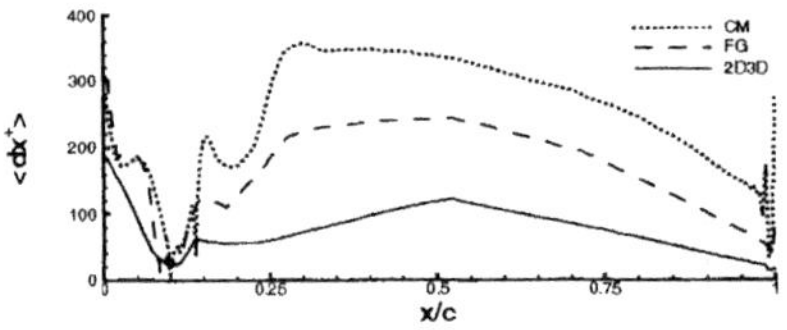

Figure 104 Karlsruhe: dx^+ vs. x/c

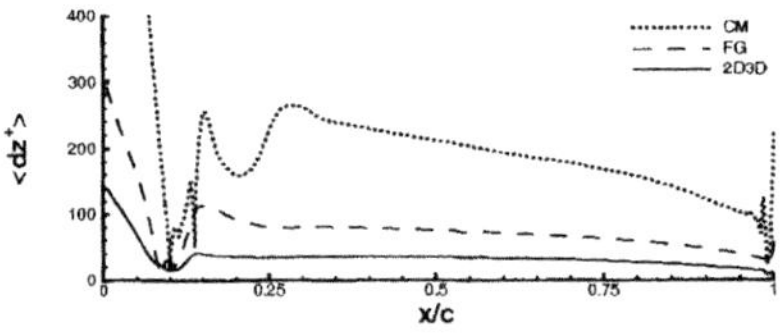

Figure 105 Karlsruhe: dz^+ vs. x/c

LES resolution requirements. The airfoil computations at Karlsruhe started on grids relatively coarse for the physics to be captured but already costly in terms of computation. Some information on these runs (CG-1 … FG-1) is compiled in the first part of Table 4.7. The results were however not satisfactory and revealed the extreme difficulty such an airfoil computation at high Reynolds number presents for LES.

In a companion study a series of channel computations were undertaken in order to understand the minimum resolution requirements for a satisfactory wall function computation of an attached turbulent boundary layer flow. The mesh used in these computations was constructed to yield similar resolution as employed in the early airfoil computations mentioned above. The near-wall cells had dimensions of $dx^+ \sim 400$, $dy^+ \sim 30$ and $dz^+ \sim 75$. The results were uniformly poor and indicated that LES computations of the airfoil using a mesh with similar resolution were unlikely to produce satisfactory outcome.

In response to these findings the Karlsruhe made a concerted effort to construct meshes with a resolution comparable to that required for an adequate wall function channel computation. Based on previous experience with the channel, nominal resolution requirements of $dx^+ < 100$ and $dz^+ < 50$ were set for the near wall cell sizing on the airfoil upper surface. Figs. 104 and 105 compare the near-wall resolution of the meshes from three of the different runs assembled in Table 4.7. The corresponding wall-normal resolution is given in Fig.123 below. Computation CM-4 used the common 'A' mesh, FG-1 used a refined mesh while the 2D3D-1 computation used an even finer mesh constructed towards the end of the project. This last mesh has dimensions of $2012 \times 72 \times 128$ grid points for a total of $18 \cdot 10^6$ cells. Approximately 1100 cells are located on the airfoil upper surface. The spanwise extent of the computational domain was set to be $L_z = 0.06c$. This mesh yields a substantially better resolution than the previous ones and largely meets the nominal resolution requirements outlined earlier. The cost of computing on this mesh is reduced using the technique to be described in the next section.

Computational Efficiency. The computational demands of Task 5 have led to two major changes to Karlsruhe LESOCC code over the course of the project. The first has been the implementation of a spanwise Fourier solver, adapting FFT routines supplied by the partner University of Surrey. In channel flow geometries the Fourier solver is up to 3 times more efficient than the original 3D-SIP solver. When applied to the airfoil the Fourier solver permitted the residuals of the p' equation to be reduced to levels unattainable with the 3D-SIP solver.

The second modification has been to introduce a 2D/3D strategy of zonal mesh refinement. This approach had been suggested by the partner ONERA in the LESFOIL project. It takes advantage of the fact that outside the airfoil boundary layer the flow is largely inviscid and two–dimensional. The fully 3D mesh then only needs to be employed in the turbulent boundary layer and the wake. Fig. 106 illustrates the concept.

The 2D mesh is in fact a single plane of 3D cells that have a z dimension equal to the spanwise extent of the 3D computational domain. Along the interface between the two zones care must be taken to ensure that mass and momentum fluxes are conserved. Careful consideration also needs to be given to the location of the 2D/3D interfaces. The 2D mesh is naturally unable to support any turbulence and hence the interfaces between 2D and 3D mesh are situated in regions which contain 2D laminar or inviscid flow. On the upper

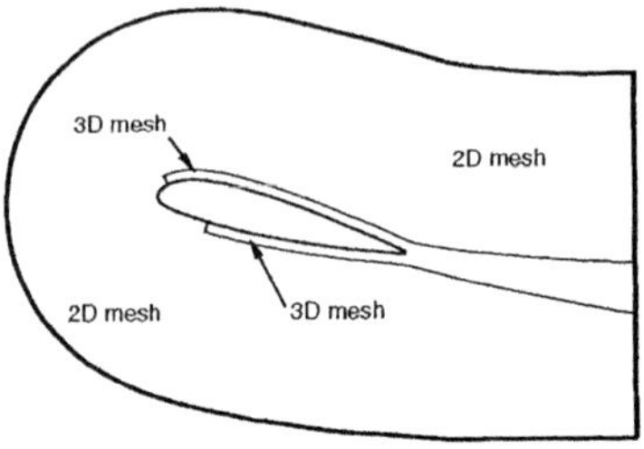

Figure 106 Karlsruhe: 2D/3D zonal mesh refinement

and lower surface the 3D mesh commences just upstream of the point of transition in the experiment ($x/c = 0.12$ on the upper surface and $x/c = 0.3$ on the lower surface). Normal to the wall the 3D mesh extends over 5-10 times the boundary layer thickness into the freestream. In the wake region the 3D mesh is continued up to the outlet from the domain.

In the computations undertaken with this strategy approximately 50% of the grid points in the 3D mesh could be removed by replacing the 3D grid with its 2D equivalent. Correspondingly the computational time required was reduced by nearly a factor of two.

Transition modelling. As an outcome of Task 3 on transition modelling it was concluded that transition would not be resolvable with the current meshes. Instead transition should be imposed by some measure at a location determined from the experimental data.

The LESOCC code employed by Karlsruhe uses second order central differencing to discretise the governing equations. Experience quickly showed that the combination of this approach with relatively coarse meshes, the high Reynolds number of the test case and correspondingly large cell Peclet numbers led to unphysical flow oscillations of considerable magnitude in the nominally laminar region around the nose of the airfoil. The approach taken for transition modelling was now to damp these spurious oscillations upstream of the transition point and to revert to the central scheme downstream of it. This was performed using the 2nd order bounded HLPA upwind scheme [77] for the evaluation of the convective fluxes in the nominally laminar regions of the flow. In the vicinity of the point of transition selected blending schemes were employed to gradually blend to the usual non-dissipative second order central differencing scheme. This concept is illustrated in Fig. 107, where blendings, conv(1) and conv(2) are shown.

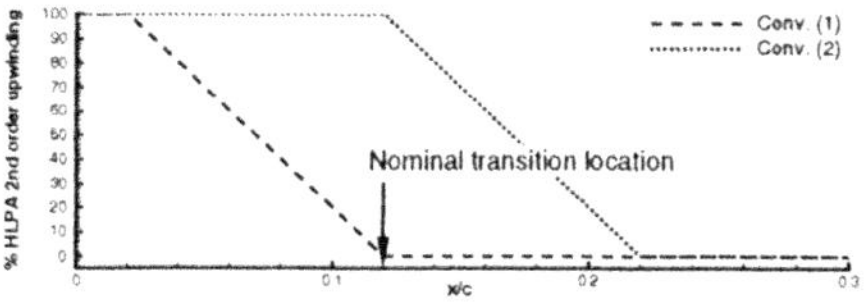

Figure 107 Karlsruhe: Convection blendings

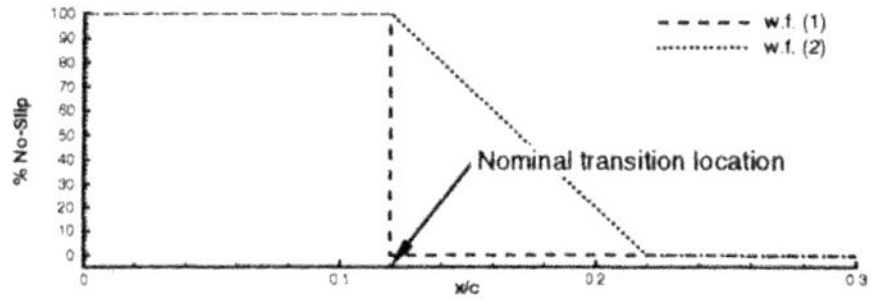

Figure 108 Karlsruhe: Wall-function blendings

At the wall the no-slip condition is applied in the nominally laminar part of the flow. Around the nominal transition location one of two different blending functions is used to gradually switch to the Werner-Wengle near-wall model. These functions, w.f.(1) and w.f.(2) are shown in Fig. 108.

Modifications to the subgrid scale model in the transition region vary. The Dynamic Smagorinsky SGS model was applied without modification. The Dynamic procedure should detect the onset of transition and act accordingly. When the plain Smagorinsky model is used ν_t was gradually applied using the chosen wall function blending.

An evaluation of the effect of the different blending schemes was performed as indicated in Table 4.7. Four different transition models were constructed using different combinations of the wall and convective blendings and then evaluated on the common 'A' mesh. These computations indicated that once the upwinding is removed the flow quickly undergoes transition and becomes turbulent. However, while the transition modelling did influence the transition process itself it was seen that the downstream development of the turbulent boundary layer was largely insensitive to the choice of model. However, it is important to note that the common 'A', mesh employed in this study had a relatively coarse streamwise and spanwise resolution, as shown previously in Figs. 104 and 105. As a result the transition process occurring in these computations was promoted nearly entirely by numerical instabilities and hence is not physically realistic. The streamwise development of the boundary layer would also have been influenced by the poor mesh resolution.

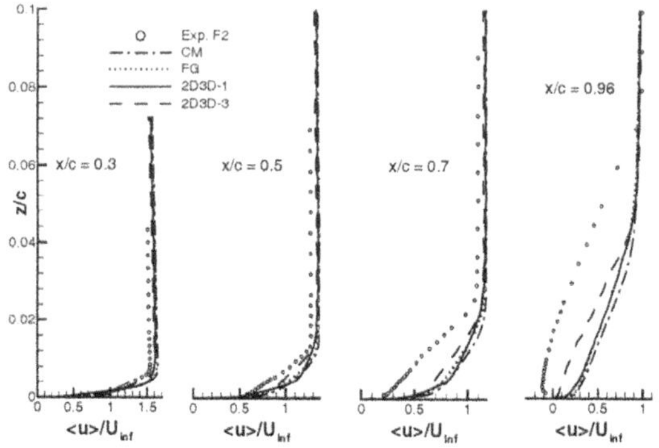

Figure 109 Karlsruhe: $\langle u \rangle / U_\infty$ profiles

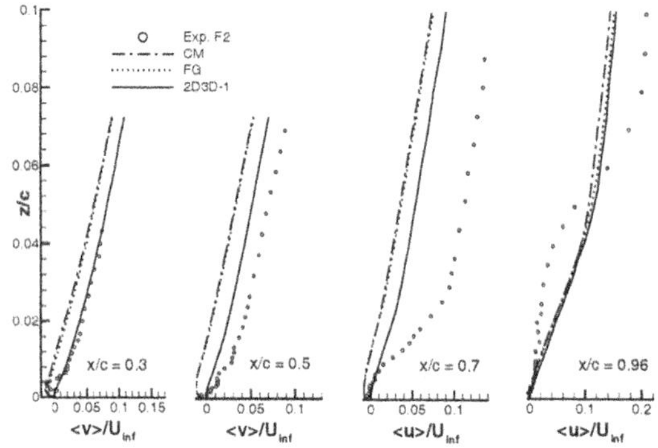

Figure 110 Karlsruhe: $\langle v \rangle / U_\infty$ profiles

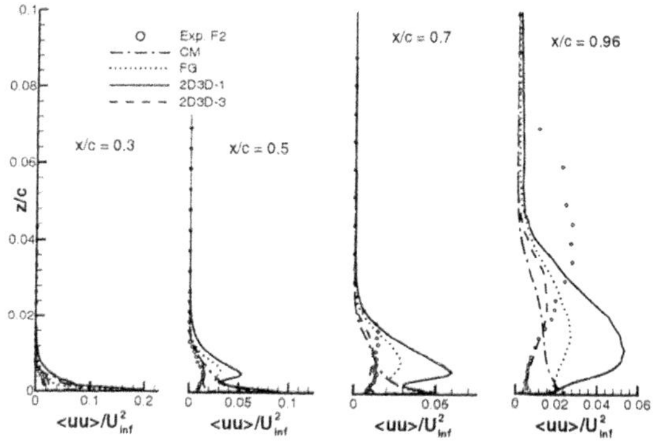

Figure 111 Karlsruhe: $\langle uu \rangle / U_\infty^2$ profiles

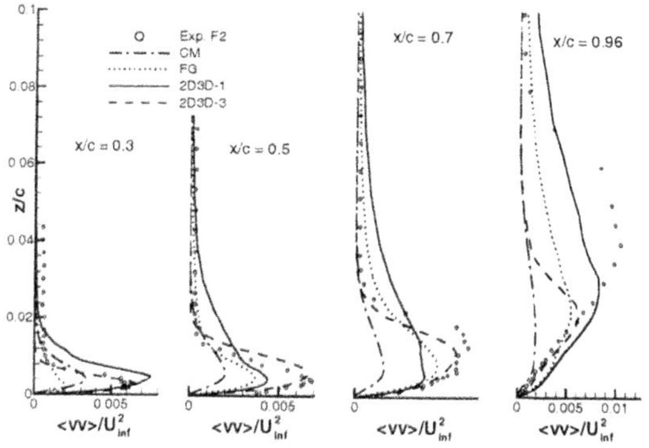

Figure 112 Karlsruhe: $\langle vv \rangle / U_\infty^2$ profiles

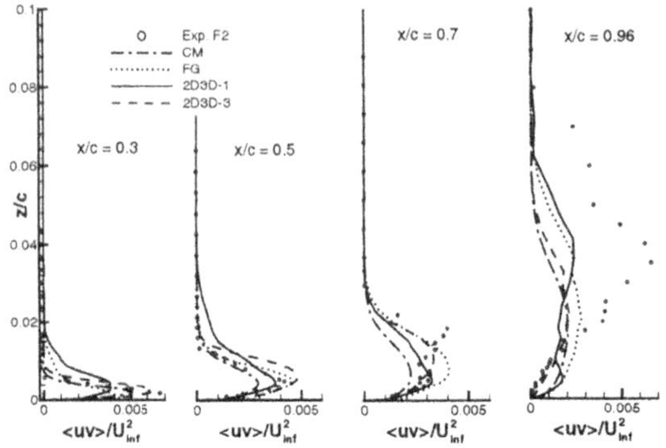

Figure 113 Karlsruhe: $\langle uv \rangle / U_\infty^2$ profiles

Time and resource constraints have not permitted a thorough examination of the transition modelling on the finer 2D/3D mesh constructed towards the end of the project. But the calculations on this mesh have shown that the transition is still induced by numerical instabilities.

On this grid, one computation included a 'transition strip' as part of the transition model. This device is intended to promote the transition process in those computations were transition does not occur spontaneously upon removal of the upwinded convection scheme. The 'transition strip' was applied at the nominal transition location x_{tr} and is constructed by modifying the applied wall shear stress τ_w in the following way :

$$\tau_w = \alpha \cdot \tau_{w,calculated} \quad \text{for} \quad x_{tr} - 0.5\%\ x/c < x < x_{tr} + 0.5\%\ x/c \qquad (19)$$

where $\alpha = 2$ in all applications to date.

Airfoil calculations. Table 4.7 contains details on the majority of the airfoil computations undertaken by Karlsruhe for Task 5. The calculations based on the 2D/3D strategy are denoted 2D3D-1 to 2D3D-3. The primary difference between these calculations and those conducted earlier in the project is the use of the high resolution mesh described in a previous section. It was hoped that this refined mesh would bring about a marked improvement in the quality of the results. Computation 2D3D-2 was undertaken at an angle of attack slightly higher than the nominal – here $\alpha = 15.3°$. The interest lay in determining if the formation of the trailing edge separation bubble could be promoted by increasing the magnitude of the suction side adverse pressure gradient. Computation 2D3D-3 differs from 2D3D-1 through the use of the transition strip described earlier. For the purposes of evaluating these results comparisons have been made with two coarse grid computations, CM-4 and FG-1 in Table 4.7. Note: In the discussion which follows these will be referred to as the CM and FG computations respectively. The evaluation which follows is performed for the suction side as this is the most relevant part of the flow.

Table 27 Karlsruhe: Summary of airfoil computations. **Note:** t.s. = transition strip, CM- ... = computation on the Common Mesh A. Two values are given for $N_{y,3D}$. These refer to number of cells in the 3D-region on the profile and in the wake (profile/wake). The reason why there are more cells in the 3D region in the wake is that the blunt edge is implemented by blocking out cells.

Comp.	**2D Mesh** ($N_x \times N_y$)	L_z	N_z	$N_{y,3D}$	**Transition blending**	**SGS model**
CG-1	516 × 94	0.13c	64	-	conv(1) + w.f(1)	Dyn. Smag - 2D filt.
CG-2	622 × 96	0.065c	64	-	conv(2) + w.f(1)	Dyn. Smag - 2D filt.
CM-1	720 × 64	0.12c	32	-	conv(1) + w.f(1)	Dyn. Smag - 2D filt.
CM-2	720 × 64	0.12c	32	-	conv(2) + w.f(1)	Dyn. Smag - 2D filt.
CM-3	720 × 64	0.12c	32	-	conv(1) + w.f(2)	Dyn. Smag - 2D filt.
CM-4(CM)	720 × 64	0.12c	32	-	conv(2) + w.f(2)	Dyn. Smag - 2D filt.
CM-5	720 × 64	0.12c	32	-	conv(2) + w.f(2)	Smag - no wall damp.
CM-6	720 × 64	0.06c	64	-	conv(2) + w.f(2)	Dyn. Smag. - 2D filt.
FG-1(FG)	1014 × 68	0.06c	64	-	conv(1) + w.f(2)	Dyn. Smag. - 3D filt.
2D3D-1	2012 × 72	0.06c	128	32/40	conv(1) + w.f(2)	Dyn. Smag. - 2D filt.
2D3D-2	2012 × 72	0.06c	128	32/40	conv(1) + w.f(2)	Dyn. Smag. - 2D filt.
2D3D-3	2012 × 72	0.06c	128	32/40	conv(1) w.f(2)+t.s.	Dyn. Smag. - 2D filt.

In all cases the Reynolds number based on chord length and freestream velocity was $Re_c = 2 \times 10^6$. The Werner Wengle law was employed to model the near-wall region and the integration time step was chosen so that the maximum CFL number was always ~ 0.65. Thus for computations CM and FG $\Delta t = 1.5 \times 10^{-4} c/U_\infty$, for 2D3D-1,3 $\Delta t = 1 \times 10^{-4} c/U_\infty$ and for 2D3D-2 $\Delta t = 7.5 \times 10^{-5} c/U_\infty$. Integration times range between 10 and 13 c/U_∞.

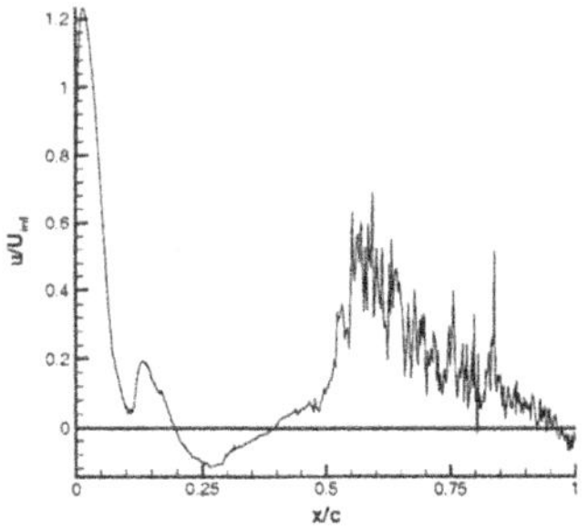

Figure 114 Karlsruhe: u/U_∞ vs. x/c at one instant on a grid line adjacent to the wall

Flow Profiles & Transition. Mean flow profiles from the CM, FG, 2D3D-1 and 2D3D-3 computations are compared with each other and to the F2 experimental data in Figs. 109 to 113. Clearly there is improvement upon grid refinement. The use of the transition strip in 2D3D-3 is also seen to be important. The trailing edge separation seen in the F2 experimental data is not captured by any of the simulations. The $\langle u \rangle / U_\infty$ profiles are everywhere too full. It is encouraging to see however that $\langle u \rangle / U_\infty$ profiles from the 2D3D-3 computation are generally the closest to the F2 results. The same is true of the resolved stress profiles. The 2D3D-3 $\langle uu \rangle / U_\infty^2$, $\langle vv \rangle / U_\infty^2$ and $\langle uv \rangle / U_\infty^2$ profiles shown in Figs. 111 to 113 are particularly encouraging, with quite good agreement between computation and experiment over much of the boundary layer.

Most of the $\langle uu \rangle / U_\infty^2$ profiles from the other computations show peaks in the near wall region. The FG and 2D3D-1 computations are also displaying a second peak about 0.01c above the airfoil surface for $x/c = 0.5, 0.7$ and 0.9. The near wall peak is not matched by the experimental data. The Karlsruhe resolution studies had clearly indicated that excessive near-wall streamwise

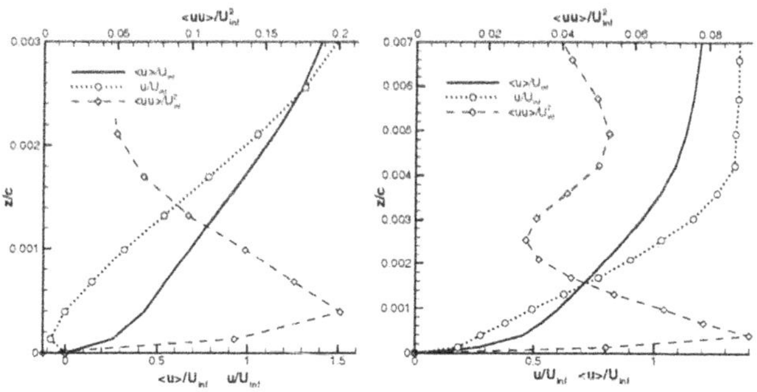

Figure 115 Karlsruhe: Instantaneous and average streamwise velocity and fluctuations in the close vicinity of the wall. **Left:** $x/c = 0.3$ **right:** $x/c = 0.5$

fluctuations could be attributed to poor grid resolution. Here the magnitude of the near-wall fluctuations appear to have increased as the resolution is improved. While the location of the second peak generally does correlate with a peak in the F2 dataset for smaller x/c, it also has a magnitude which is too large.

Insight into the 2D3D-1 $\langle uu \rangle / U_\infty^2$ profiles is given by Figs. 114 to 115. Fig. 114 gives a plot of the u velocity at the first grid point adjacent to the wall for one instant in time. It is seen that at this instant the near-wall flow is very much laminar until approximately $x/c = 0.55$. Moreover a laminar separation bubble stretches between $x/c \sim 0.2$ and $x/c \sim 0.5$. This large bubble is only a transient phenomenon. At a slightly later instant in time transition occurs at a position close to the experimentally indicated location of $x/c \sim 0.12$ and the turbulent boundary layer reattaches. The bubble which has formed persists however and is convected downstream. The formation and shedding of this bubble is cyclical with a period close to c/U_∞. Fig. 115 shows details of the computation 2D3D-1. For the same instant as Fig. 114 it displays the instantaneous and the average velocity as well as the stream wise fluctuation at $x/c = 0.3$ and $x/c = 0.5$. It is seen that the separation typically has a wall-normal depth of only 1-3 cells and it is thus extremely shallow and poorly resolved. Additionally, due to the region of reverse flow the separated laminar boundary layer deviates markedly from the mean profile – especially near the wall and close to $z/c = 0.05$. At this instant the correlation between the location of maximum deviation and the peaks in the resolved $\langle uu \rangle / U_\infty^2$ stress levels appears to be strong.

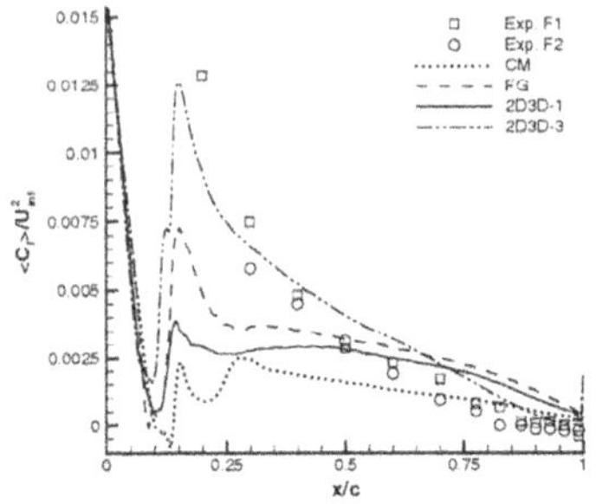

Figure 116 Karlsruhe: $\langle C_f \rangle / U_\infty^2$ vs. x/c

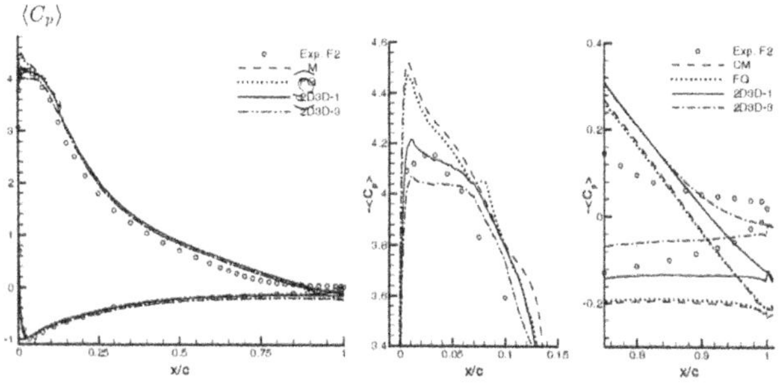

Figure 117 Karlsruhe: Pressure coefficient $\langle C_p \rangle$. **Left:** entire profile, **middle:** zoom around nose **right:** zoom around trailing edge.

In the experiments, a small laminar separation bubble was observed just before transition at $x/c \sim 0.12$. The computational bubble often originates from this location and is hence essentially physically correct. However it is possible that inadequate wall normal resolution of the mesh at some instances inhibits correct transition of the separated laminar boundary layer. The laminar separated region then grows in extent and sheds only after transition and once again initiates at or near the correct location of $x/c \sim 0.12$

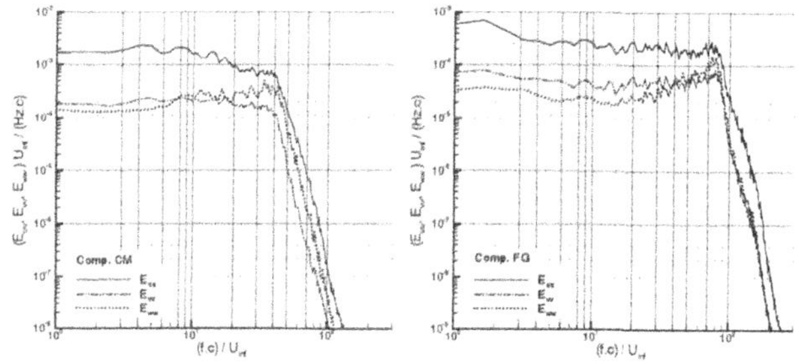

Figure 118 Karlsruhe: u, v, w spectra $x/c = 0.3$

In computation 2D3D-3 the formation and shedding of this physically unrealistic bubble was prevented by the addition of a 'transition strip' at $x/c = 0.12$, the form of which was described in an earlier section. This numerical perturbation ensured that transition always occured at the desired location. The results from this computation are far superior to those from 2D3D-1. The mean streamwise velocity profiles $\langle u \rangle / U_\infty$ are much closer to separation and the large near-wall peaks in $\langle uu \rangle / U_\infty^2$ are no longer present. The results from this computation might have been further improved if the 2D-3D computational boundaries had been better placed. At $x/c = 0.96$ it is noticeable that velocity fluctuations are largely absent for $y/c \sim 0.05$ and above. In fact the region above $y/c = 0.05$ is a 2D computational domain and thus it seems possible that the development of the separated region has been constrained.

Fig. 116 shows the suction side skin friction coefficient C_f returned by the three computations. The 2D3D-3 computation comes closest to returning the experimental behaviour. However no computation correctly captures the large peak in the experimental C_f just after transition while near the trailing edge C_f is uniformly overpredicted. This is related to the fact that all computations failed to capture the trailing edge separation bubble.

Table 28 Karlsruhe: Comparisons of C_L and C_D with experiments F1 and F2.

Result	CM	FG	2D3D-1	2D3D-3	Exp. F1	Exp. F2
C_L	1.71	1.69	1.64	1.54	1.56	1.515
C_D	0.0158	0.0168	0.0170	0.0223	0.0204	0.0308

Lift and Drag. Table 4.7 shows mean C_L and C_D for computations CM, FG, 2D3D-1 and 2D3D-3. Generally C_L is overpredicted while C_D is underpredicted. The 2D3D-3 computation gives the best lift and drag estimates. Fig. 117 shows the pressure distribution for the different computations. It is clear that the 2D/3D computations best capture the leading edge suction peak and also come closest to correctly predicting the pressure distribution at the trailing edge. Generally the correct pressure distribution about the airfoil will not be achieved until the trailing edge separation is captured. From observing the pressure distributions shown in Fig. 117 it does appear that encouraging progress is being made towards achieving this goal.

Velocity/Pressure Spectra and Correlations. Knowledge of the velocity and pressure spectra of the flow about an airfoil can give information which may be useful for both aero-elasticity and noise production analysis. The spectra can also provide important insight into the adequacy of the resolution of an LES computation. Figs. 118 to 119 show the resolved power spectral densities of the u, v and w velocity components taken from computations CM and FG at selected locations on the suction side of the airfoil. In all cases the flow was sampled at a wall normal distance of about $y^+ = 150$.

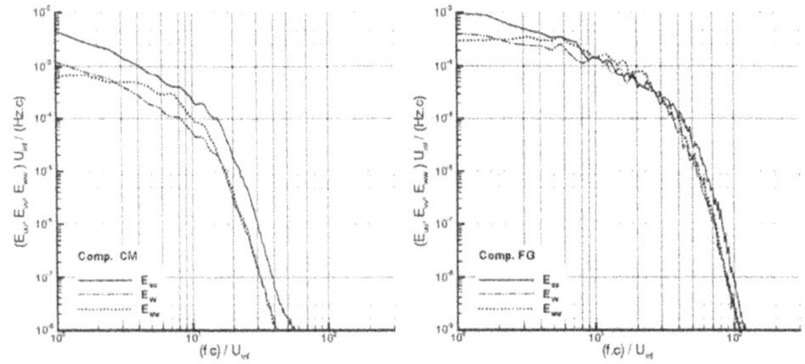

Figure 119 Karlsruhe: u, v, w spectra $x/c = 0.96$

At $x/c = 0.3$ the spectra from both computations have a shape which is characteristically seen in computations with inadequate resolution. The energy drain from the small scales is not correctly captured and as a result there is an energy pile-up near the cutoff frequency. This pile-up is most noticeable in the spanwise fluctuations and even more pronounced for the finer grid. At this location Figs. 104 and 105 show that the spanwise and streamwise resolution of the respective meshes are: CM $\Delta x^+ = 360$, $\Delta z^+ = 260$; FG $\Delta x^+ = 223$, $\Delta z^+ = 80$. The spectra from $x/c = 0.96$ (Fig. 119) show no such pile-up and have a shape which is much closer to what is typical for a turbulent flow. As expected the spectrum from the FG computation has a cutoff frequency which is appreciably higher than that of the CM computation. At this location the respective mesh sizes are: CM $\Delta x^+ = 140$, $\Delta z^+ = 100$; FG $\Delta x^+ = 70$, $\Delta z^+ = 40$.

Spectra from the 2D/3D mesh are not available. However the resolution of this mesh is such that $\Delta x^+ < 100$ and $\Delta z^+ < 40$ over nearly the entire suction surface. The spectral quality of any LES conducted on this mesh should therefore everywhere equal or exceed that of the $x/c = 0.96$ spectra shown above.

Figs. 120 to 121 show two-point spanwise velocity correlations of u, v and w in the CM and FG computations at $x/c = 0.3$ and $x/c = 0.96$, respectively. They have been obtained from time signals at several points shifted in spanwise direction so that the resolution of the correlation plot is coarser than that of the computational grid. In the FG computation the u and v correlation levels are substantial. Interestingly, these levels do not appear to decay as the separation Δz increases. This indicates the presence of substantially coherent 2D flow structures extended across the entire spanwise extent of the domain. It also suggests that the spanwise extent of the computational domain was perhaps insufficient. Here $L_z = 0.06c$ with 64 computational cells in the spanwise direction. In contrast the CM computation, with $L_z = 0.12c$, generally displays very low correlation levels.

The impact of a substantial non-zero spanwise correlation on the results is not clear. However achieving a near-zero spanwise correlation while still maintaining a reasonable spanwise mesh resolution to date is prohibitively expensive. Generally the reduced spanwise extent of the FG and 2D/3D meshes (both with $L_z = 0.06c$) has been employed in order to enable a reasonable spanwise mesh resolution to be achieved for a minimum number of spanwise cells. The 2D/3D mesh, with $N_z = 128$, achieves everywhere $\Delta z^+_{max} \sim 40$. Doubling the spanwise extent of this mesh while maintaining the spanwise resolution would require $N_z = 256$ and yield $36 \cdot 10^6$ cells in total. Even the 2D/3D zonal approach described earlier would require approximately $18 \cdot 10^6$ computational cells for this mesh. With these figures in mind the current strategy has been to achieve a reasonable spanwise resolution in preference to a near-zero spanwise correlation. The issues of resolution, transition etc.

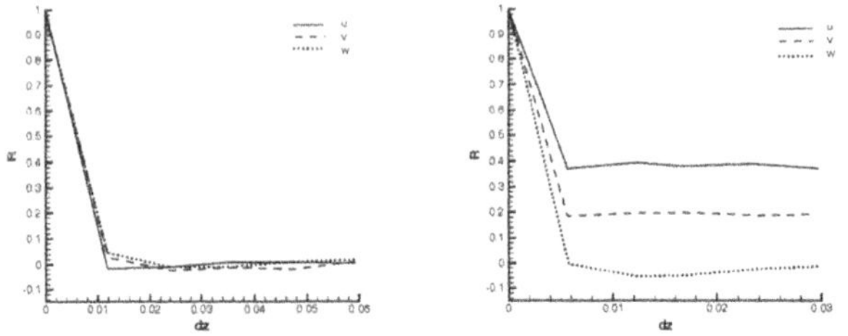

Figure 120 Karlsruhe: Spanwise correlations of u, v, w at $x/c = 0.3$. **Left:** computation on the grid CM, **right:** computation on the grid FG.

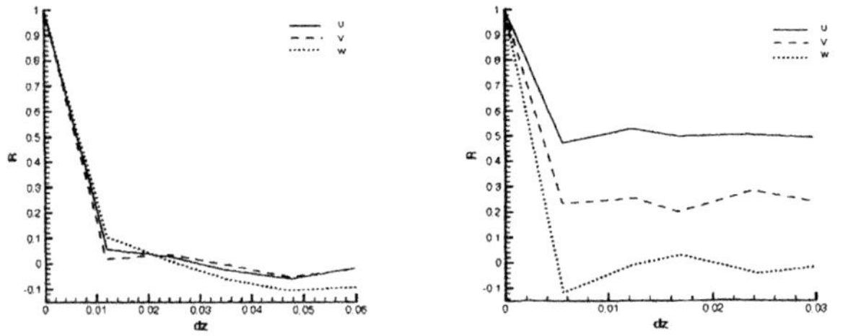

Figure 121 Karlsruhe: Spanwise correlations of u, v, w at $x/c = 0.96$. **Left:** computation on the grid CM, **right:** computation on the grid FG.

are largely independent of the spanwise extent and can adequately be investigated in this setting.

SGS Modelling. Fig. 122 shows the ratio of eddy viscosity to molecular viscosity ν_t/ν at different locations along the suction side of the airfoil. In all cases the Dynamic Smagorinsky model has been used to provide the SGS stresses. The dynamic procedure employed a two dimensional filter in $x-z$ and used spanwise and temporal averaging to regularise the model coefficient.

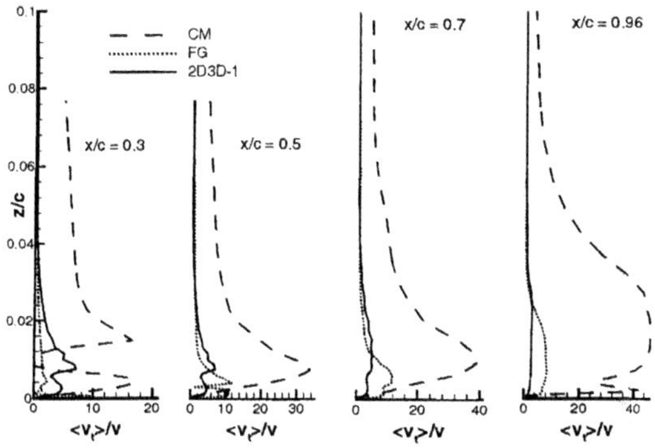

Figure 122 Karlsruhe: $\langle\nu_t\rangle/\nu$ profiles

The impact of the mesh refinement is clear. The contribution of the SGS model decreases as the levels of the resolved motions increase. All the profiles have been obtained after sampling for at least 10 c/U_∞ corresponding to approximately 100000 time steps for most of the computations. Despite this long averaging it is noticeable that the upstream profiles from the FG and 2D3D-1 computations are not particularly smooth. For the 2D/3D computation this observation corresponds well with the low frequency modulations described earlier.

Wall Modelling. During Task 5 it has been the policy of Karlsruhe to concentrate on the use of wall function meshes. In conjunction with this the Werner-Wengle model has been used exclusively to model the near-wall region. Karlsruhe experience with applying the Werner-Wengle near-wall model to channel computations as reported in Task 2 has shown that best results are achieved when the first grid point from the wall is located at $30 \leq y_1^+ \leq 60$. Fig. 123 shows y_1^+ for the three airfoil meshes described in this report.

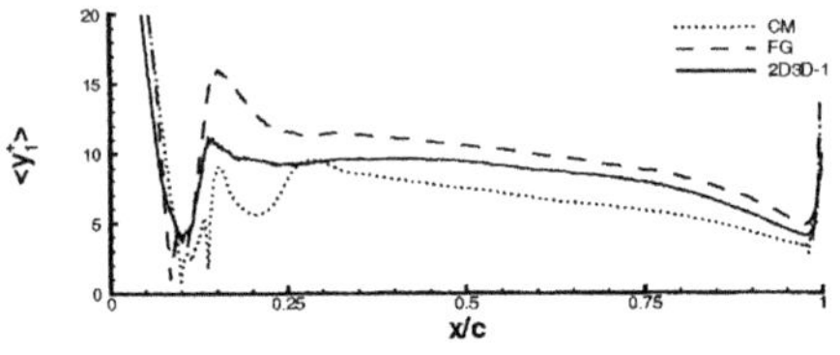

Figure 123 Karlsruhe: y_1^+ vs. x/c

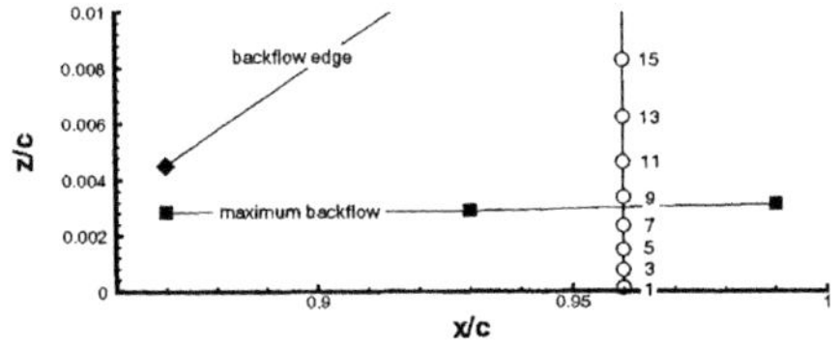

Figure 124 Karlsruhe: Detail of reverse flow region

For all these calculations y_1^+ is nearly everywhere 10 or below. When y_1^+ is in this range the Werner-Wengle model almost acts like a no-slip condition. The difference with respect to the latter is, however, that locally and instantaneously the value of y_1^+ can become large such that the 1/7-power law is applied. This effect is relatively small so that the choice of wall model has largely been irrelevant to the outcome of the computation. The choice of y_1^+ is dictated by the necessity of having a wall-normal resolution which permits at least marginal resolution of the reverse flow region of the trailing edge separation bubble, specifically the region below the point of maximum backflow. Fig. 124 indicates the position of maximum backflow above the wall in the experiments. For evaluation selected points of the grid employed in the 2D/3D computations have been inserted at $x/c = 0.96$. There are approximately 7 cell centres inside the maximum backflow region. However, if the mesh were to be altered so that y_1^+ is 30 or more then there would be at most one cell center below the point of maximum backflow. Hence, the position of the wall-adjacent point is dictated by the necessity to resolve the recirculating flow.

The extent to which the near-wall modelling has been degraded by these compromises is not clear. Certainly y_1^+ is inside the range where the Werner-Wengle wall model is able to function effectively. However y_1^+ is also much larger than what is usually considered necessary for the application of the no-slip condition. These concerns can also be viewed in the light of the somewhat unsatisfactory C_f profiles shown in Fig. 116 and the under-resolved laminar separation bubble described in an earlier section.

Computation 2D3D-2: $\alpha = 15.3^o$. A computation at a higher angle of attack was undertaken as an attempt to gauge the sensitivity of the trailing edge separation bubble to the magnitude of the suction side adverse pressure gradient. This line of investigation was suggested by inconsistencies between the F1 and F2 experimental data sets. While the F2 data suggests the onset of stall at around 13.3°, the F1 data seems to indicate a slightly larger angle of attack, close to $\alpha = 15°$.

The results of the 2D3D-2 computations at $\alpha = 15.3^o$ are only slightly different from those of the 2D3D-1 run. In particular they also do not show a trailing edge separation bubble. The flow does respond favourably to the increased pressure gradient as the 2D3D-2 $\langle u\rangle/U_\infty$ profiles are not quite as full as those from 2D3D-1. It is thus likely that the computation will display separation for a large enough value of α. Importantly though it can be concluded that the lack of separation is not a result of choosing an inappropriate angle of attack.

The results from computation 2D3D-3 suggest that the 2D3D-2 results would have been improved if a transition strip had been included.

Trailing Edge Geometry. Initially, most partners had performed airfoil computations with a sharp trailing edge as in all the RANS computations of the literature [2]. The experiments, however were conducted with a blunt trailing edge of height $h_e = 0.05c$. During a later stage of the project Dassault displayed results incorporating such a blunt edge and suggested that this feature had a significant impact on the flow around the airfoil.

All 2D/3D computations presented in this report have included this geometrical feature. On the other hand the resolution was substantially improved with respect to previous computations which also has an impact as discussed above. Thus, while the importance of the correct trailing edge geometry cannot be discounted, the presence of the blunt edge by itself seems not sufficient to guarantee that separation will occur.

Conclusions. The Task 5 effort has seen Karlsruhe complete over a dozen separate airfoil calculations. It became apparent that the actual choice of a SGS model is presumably not as decisive for the outcome of the simulation as the resolution. Hence, classical SGS models have been employed. For the wall modelling, wall functions have been used which proved successful in attached turbulent boundary layers as previously investigated in Task 2. Here, however, deficiencies were experienced due to the resolution as was discussed.

One of the important outcomes from the Karlsruhe work has been that a truly successful LES of the airfoil flow is unlikely unless the mesh employed meets recognised LES resolution requirements. The growth in the number of grid points during the course of the project reflects this fact: early computations employed $\sim 1.5 \times 10^6$ cells whereas final computations used meshes with $\sim 18 \times 10^6$ cells increasing the cost over-proportionally due to the resulting reduction of the time steps.

For computational cost reasons the issue of the spanwise extent of the calculation domain is one which is closely tied to the question of mesh resolution. Too small an extent results in excessive spanwise velocity correlations and introduces concern as to the physical realism of the results. This should be avoided if at all possible. Doing so while maintaining adequate cell resolution in the spanwise direction can however add significantly to the cost. Hence, a compromise is neccessary. To perform cost-effective LES, Karlsruhe has worked intensively to improve the efficiency of the employed numerical method. Improvements in the numerics involved the implementation of a highly efficient Fourier solver for the p' equation and an introduction of a technique of zonal 2D/3D mesh refinement. This in fact is a variant of a local grid refinement with non-matching cells. It would be beneficial for the cost of such simulations to further develop and generalize this approach in the code employed so as to become even more flexible and economic with block structured grids.

Extensive work has also been carried out on the question of transition modelling and methods for controlling the transition process have been successfully developed. However, even on the finest mesh used the transition which occurs is not physically realistic but numerically triggered. If the departure from physical realism is too large, however, successful simulation downstream of this region is unlikely. In the case considered, transition takes place in and after a laminar separation bubble near the airfoil leading edge. In the computations this bubble was not well represented and it is likely that this was determinental to the results. Work on this issue is currently pursued outside the project.

4.8 Contribution by ONERA

Authors: I. Mary and P. Sagaut

Introduction. The ONERA contribution to the LESFOIL project [78] has seen the introduction of new numerical techniques designed to enhance the capability of LES to predict the flow around an isolated airfoil. These include:

- The development of a 2D/3D coupling strategy in a structured solver, FLU3M, in order to reduce the total number of grid points. In the airfoil simulations the flow is laminar and two-dimensional in a large part of the computational domain. Thus the three-dimensional computation can be limited to a zone close to the airfoil where the flow is turbulent and two-dimensional simulations are used for the far field (see Fig.125). This strategy, which does not alter the accuracy of the simulation, allows the use of a mesh with LES resolution while keeping the total number of grid points at a reasonable level.
- The development of a second-order accurate hybrid centered/upwinding formula for the Euler fluxes. The key point of the scheme is the use of a sensor which permits the local introduction of some numerical dissipation when a numerical wiggle is detected on one of the primitive variables. Therefore, the effect of the SGS model is not affected by the numerical dissipation as long as odd-even decoupling is not detected in the flow solution.

Details on these are given in the sections which follow.

Simulation method. The solver FLU3M, developed by ONERA, is based on the compressible filtered Navier-Stokes equations. The discretization is realized on multi-block structured meshes using a finite volume technique. A second-order accurate implicit temporal integration is achieved thanks to an approximate Newton method. General information about the solver is available in Refs.[79, 80].

Subgrid Scale Modelling. Due to the large computational cost of the intended simulation only one explicit subgrid scale model has been used for this study. The selective mixed scale model (SMSM), developed by Sagaut and Lenormand, has been employed because it realizes a good compromise between accuracy, stability and computational cost [81, 82]. More particularly, the use of a selective function allows one to handle transitional flows [83] which is one of the key points of the present application.

The eddy viscosity is given using a non-linear combination of the filtered shear stress tensor, a characteristic length scale, the small scale kinetic energy, and the selective function:

$$\mu_t = \bar{\rho} f_{\theta_0}(\theta) C_m \Delta \sqrt{0.5 S_{ij}(\tilde{u}) S_{ij}(\tilde{u})} \sqrt{K_c} \quad (20)$$

where the test field kinetic energy is evaluated as

$$K_c = \sqrt{0.5 (\tilde{u}_i)' (\tilde{u}_i)'} \quad (21)$$

The test field $(\tilde{u}_i)'$ is extracted from the resolved field, $(\tilde{u}_i)' = \tilde{u}_i - \widehat{\tilde{u}}_i$, employing an averaging test filter denoted by a wide hat:

$$\widehat{\tilde{u}}_i = 0.25 \tilde{u}_{i-1} + 0.5 \tilde{u}_i + 0.25 \tilde{u}_{i+1} \quad (22)$$

The characteristic length scale, Δ, is given by the cell volume of the mesh, and the constant parameter, $C_m = 0.06$, is derived in Ref.[84]. The selective function is defined by:

$$f_{\theta_0}(\theta) = \begin{cases} 1 & \text{if } \theta > \theta_0 = 10^o \\ tan^4(\frac{\theta}{2})/tan^4(\frac{\theta_0}{2}) & \text{otherwise} \end{cases} \quad (23)$$

where θ is the angle between the filtered vorticity, $\omega(\tilde{u})$ and the local averaged filtered vorticity, $\omega(\widehat{\tilde{u}})$.

A simulation has also been carried out with the MiLES approach [85] which does not explicitly use a subgrid scale model based on physical assumptions. Here the intrinsic dissipation of the numerical scheme is assumed to transfer the energy from the large scales to the small ones.

Euler Flux Discretization. The AUSM+(P) [86,87] scheme constitutes the basis of the discretization because it is well adapted to low Mach number boundary layer simulations. However, several modifications have been introduced to enhance its accuracy and reduce its computational cost. As the shock capturing properties of the scheme is not relevant for the present application, a simplified formula was developed to approximate the Euler face fluxes, in this case:

$$\bar{F}_1^{i+1/2} = U_1 \frac{Q_L + Q_R}{2} - |U_{dis}| \frac{Q_R - Q_L}{2} + \frac{\bar{p}_L + \bar{p}_R}{2} \quad (24)$$

where U_1 denotes the interface fluid velocity, L/R the left and right third-order MUSCL interpolation. The state vector Q is defined as $(\bar{\rho}, \bar{\rho}\tilde{u}_1, \bar{\rho}\tilde{u}_2, \bar{\rho}\tilde{u}_3, \widehat{\rho E} + \bar{p})^t$, whereas U_{dis} characterizes the numerical dissipation acting on the velocity components. To enforce the pressure/velocity coupling in low Mach number zones a pressure stabilization term is added to the interface fluid velocity, as done by Rhie and Chow [47] for incompressible flow:

$$U_1 = \frac{\tilde{u}_{1\,L} + \tilde{u}_{1\,R}}{2} - c_2(\bar{p}_R - \bar{p}_L) \qquad (25)$$

where c_2 is a constant parameter. The numerical dissipation U_{dis} is defined as follows:

$$U_{dis} = \Phi \max\left(\frac{|\tilde{u}_{1\,L} + \tilde{u}_{1\,R}|}{2}, c_1\right) \qquad (26)$$

where c_1 is a constant parameter and Φ a sensor used to minimize the numerical dissipation. To achieve a good compromise between stability and accuracy the constant parameters c_1 and c_2 are both set equal to 0.04 [88]. In the same spirit as the modification of the Jameson scheme proposed by Ducros [89], the present sensor Φ is a binary function which depends only on the smoothness of the primitive variables $\psi = (\bar{\rho}, \tilde{u}_1, \tilde{u}_2, \tilde{u}_3, \bar{p})^t$. If no numerical wiggle is detected on ψ at cell i, Φ is equal to 0. Otherwise Φ is set to 1 at the $i + 1/2$ and $i - 1/2$ interfaces. In order to detect a numerical wiggle the following functions are introduced:

$$\Delta^i_\phi = \begin{cases} -1 \text{ if } (\phi_{i+2} - \phi_{i+1})(\phi_{i+1} - \phi_i) < 0 \\ 1 \text{ else} \end{cases} \qquad (27)$$

$$W_{\psi_k} = \begin{cases} 1 \text{ if } \Delta^i_{\psi_k} + \Delta^{i+1}_{\psi_k} < 0 \text{ or } \Delta^i_{\psi_k} + \Delta^{i-1}_{\psi_k} < 0 \\ 0 \text{ else} \end{cases} \qquad (28)$$

These functions define the sensor in Eq. 26 as follows

$$\Phi = \max(W_{\psi_k}) \quad \text{for k=1,5} \qquad (29)$$

However if the MiLES approach is retained as the SGS "model" the sensor is set to 1 independently of the smoothness of the primitive variables. In such a case the energy transfer from the large scales to the small ones is achieved by the second term of Eq. 24.

2D/3D coupling method. As LES of wall bounded flow requires very small cells near the wall it becomes essential to optimize the cell distribution in order to limit the total number of grid points. The specificity of external flows around infinite span bodies consists in the existence of a large zone where the flow is laminar and two-dimensional. Thus the three-dimensional computation can be limited to a zone close to the airfoil where the flow is

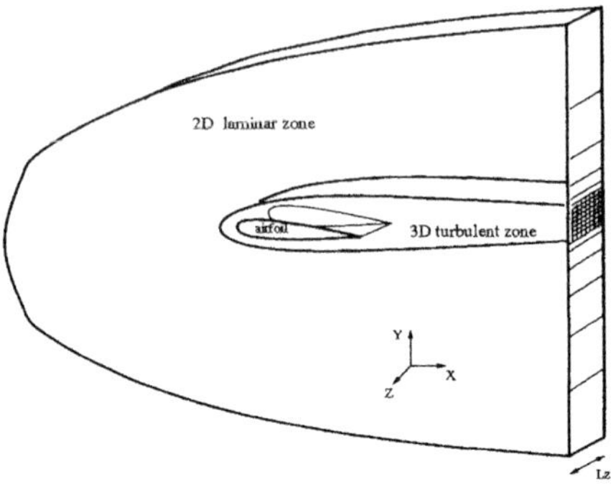

Figure 125 ONERA: 2D/3D coupling strategy.

turbulent whereas two-dimensional laminar simulations are used for the far field (see Fig.125).

As boundary conditions are treated through the use of ghost cells the 2D/3D coupling is realized by the following definitions of ghost cell data:

$$Q_c^{2D}(x,y) = \frac{1}{L_z}\int_0^{L_z} Q_c^{3D}(x,y,z)dz \qquad (30)$$

$$Q_c^{3D}(x,y,z) = Q_c^{2D}(x,y) \qquad (31)$$

where L_z denotes the spanwise extent of the computational domain. If the boundary between 2D and 3D blocks is located too close to turbulent zones the flow variables in the 3D block are subject to an implicit spanwise filtering due to Eqs.(30-31). Therefore the extent of the 3D zone in the direction normal to the wall must be chosen carefully to realize a good compromise between efficiency and accuracy. For flow over an airfoil the turbulent boundary layer thickness, which grows from the leading edge to the trailing edge, is often less than 10 percent of the chord. Thus the 2D/3D technique can easily reduce the total number of grid points by a factor of 2, especially if several 3D blocks with different extents in the direction normal to the wall are used to fit the turbulent boundary layer topology.

Computational Setup. Four C-H meshes have been used to evaluate the simulation sensitivity to different spatial resolution parameters. In the (x,y)

plan (see Fig.125) all meshes have a computational domain limit located at approximately ten chord lengths remote from the airfoil. The main characteristics of the meshes are summarized in Table 4.8.

Table 29 ONERA: Mesh characteristics

Mesh	Wake	PS	SS	N_2	N_3	L_z/c	$2D/3D$	$Nb.pts$
M1	96	430	2048	90	30	1.2%	yes	3.2×10^6
M2	96	430	2048	90	30	0.5%	yes	3.2×10^6
M3	96	430	2048	90	72	1.2%	yes	7.2×10^6
M4	57	192	192	64	32	3%	no	10^6

The rows Wake, PS, SS, N_2, N_3 and $Nb.pts$ correspond to the number of cells in the wake, on the pressure side, on the suction side, in the direction normal to the wall, in the spanwise direction and the total number of cells, respectively.

For the M4 mesh, which is identical to the mesh of [56], the blunt trailing edge is replaced by a sharp one for sake of simplicity. Even if the wall normal resolution is sufficient ($\Delta y^+ \approx 2$) the average streamwise and spanwise resolution on the suction side is very crude for LES ($\Delta x^+ \approx 800$ and $\Delta z^+ \approx 200$, respectively). This mesh hence allows to illustrate the behaviour of strongly under-resolved LES for this flow.

The M1,..,M3 grids have been designed to obtain a classical LES resolution. In the (x, y) plan, all meshes are identical. The blunt trailing edge is discretized through a $96 \times 40 \times N_3$ H-H block. Previous LES [56,78] and simulations on the M4 mesh have shown that the pressure side boundary layer remains laminar due to the favorable pressure gradient while in the experiment transition was tripped at $x/c = 0.3$. Thus for the M1,..,M3 grids no attempt has been carried out to trigger the transition on the pressure side. Therefore only 430 cells are located on this side and a 2D block is used from $x/c = 0.01$ to $x/c = 0.98$ to limit the number of points (see Fig.126). In the same way two 3D blocks, with wall normal extent close to 0.015c and 0.1c respectively, are used on the suction side to fit the growth of the boundary layer. The first block, with 38 cells in the wall normal direction, is located between $x/c = 0.01$ and $x/c = 0.38$. The second one goes from $x/c = 0.38$ to the limit of the computational domain, with 48 cells in the wall normal direction.

The grids M1, M2, M3 only differ by their spanwise characteristics. On the one hand the M2 and M3 meshes, with spanwise extents of 0.012c and 0.005c respectively, allow to check the influence of this parameter since their spanwise resolutions are equal. On the other hand the M1 and M3 meshes highlight the effect of the spanwise resolution. Fig. 127 displays the resolution of M3 in wall units (This unit is evaluated with the mean skin friction coef-

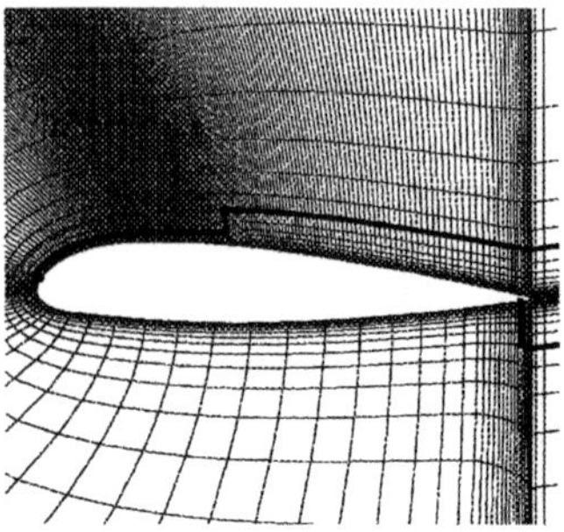

Figure 126 ONERA: Mk (k=1,3) mesh in the (x, y) plan (10% of the nodes is represented in the streamwise direction and 50% in the wall normal direction). The thick line represents the limit between 2D and 3D zones.

ficient of the M3 simulation which is in good agreement with experimental measurement as illustrated by Fig. 131). The mesh resolution usually required for LES of wall bounded flows ($\Delta y^+ \approx 2$, $\Delta z^+ \approx 20$, $\Delta x^+ \approx 100$) is almost reached everywhere.

The problem with these grids is related to the limited spanwise extent ($\leq 0.012c$), which could happen to be too small near the trailing edge. However available supercomputers and the current numerical method do not allow to handle larger domains while maintaining LES resolution.

All simulations were obtained by using no-slip conditions at the wall and periodic conditions in the spanwise direction. Non-reflecting characteristic boundary conditions were applied at the far field boundaries. Additionally, RANS simulation using the Baldwin-Lomax model provided an initial flow solution. The time step was fixed to $\Delta t = 3.3 \times 10^{-5} c/U_\infty$, implying a maximal CFL number of about 2, based on the free-stream velocity. Since the computations were performed at a Mach number of 0.15 this yields a CFL number based on the sound speed of about 16. For the M1 simulation around 6 time units c/U_∞ are necessary to get a pseudo periodic signal starting from the RANS solution. To reduce the CPU time the M1 flow field was retained as initial condition for the M2 and the M3 simulation, limiting the initial transient to 1.5 c/U_∞. Averaging was performed in the homogeneous spanwise direction and in time over a period of 2.4 c/U_∞. The simulations

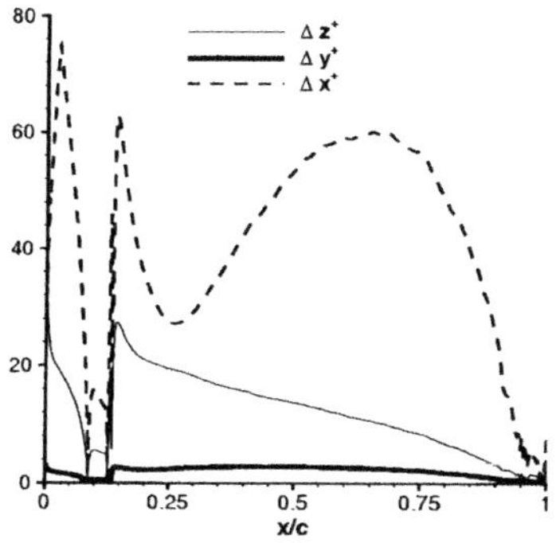

Figure 127 ONERA: M3 mesh: Evolution of the mesh resolution on the suction side in wall unit.

were carried out on a single processor of the NEC SX5 with the code running approximately at 4Gflops (for a crest speed of 8Gflops). Almost 366 CPU hours were needed to perform the time integration over a period of 2.4 c/U_∞ on the largest grid M3.

Results and Discussion In order to discard inaccurate results Fig.'s 128-129 depict the mean pressure coefficient $C_p = 2(p - p_\infty)/\rho U_\infty^2$ and the mean skin friction coefficient $C_f = 2\tau_w/\rho U_\infty^2$ (τ_w represents the wall shear stress) obtained with the different SGS modelling on the M1 and the M4 grid. It appears clearly from Fig. 128 that the mesh resolution in the (x, y) plan has a deeper impact on the solution accuracy of C_p than the SGS modelling. Indeed, the MiLES and SMSM results are nearly identical on both meshes, whereas the M1 and M4 results strongly differ. Results associated to the M4 meshes are characterized by a strong overestimation of the suction peak value and no plateau is detected near the trailing edge, indicating that no separation occurs. The C_p-distributions on M1 mesh are in very good agreement with experimental data and more particularly the SMSM results. Indeed, both the plateau and suction peak values are well predicted, whereas the sharp gradient near $x/c = 0.12$ underlines the presence of laminar separation bubble and turbulent reattachment, as mentioned in the experimental report [90].

Concerning the MiLES solution, the C_p-distribution begins to slightly differ from experimental data at $x/c = 0.8$ since no plateau is detected. This

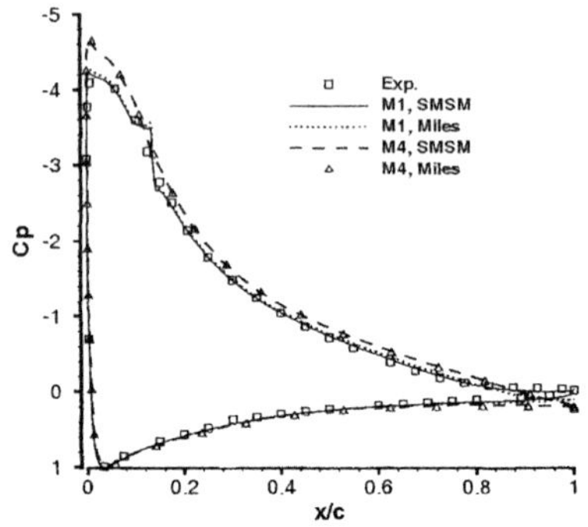

Figure 128 ONERA: Effect of SGS modelling and (x,y) mesh resolution on the mean pressure coefficient C_p.

indicates that no separation occurs with the implicit SGS modelling, which is confirmed by the value of the skin friction coefficient on Fig. 129. As this variable is strongly related to viscous effects both the SGS modelling and the mesh resolution have a significant effect on the suction side C_f-distribution. The M4 simulations lead to earlier transition than those employing M1, independently of the SGS modelling, but both models yield unsatisfactory behaviour of the C_f-distribution. Therefore, for reasons of brevity the analysis presented hereafter is mainly focused on the M1,..,M3 solutions obtained with the SMSM. These are the only calculations predicting separation near the trailing edge.

As the C_p-distributions on the M1,..,M3 meshes are almost identical between $x/c = 0$ and $x/c = 0.6$ on the both sides of the airfoil Fig. 130 shows only the influence of the spanwise properties of the mesh near the trailing edge. The size and intensity of the separation zone can be estimated by the value and extent of the plateau. As expected, it appears that both the spanwise extent and resolution have a significant impact on this flow region. Indeed, the M3 simulation exhibits a very good agreement with experimental data while the use of a smaller spanwise extent in the M2 simulation leads to a large overestimation of both the size and the intensity of the separation. The use of a coarser spanwise resolution in the M1 simulation implies a too weak separation. It seems that the reduced spanwise extent of the M2 mesh leads to a nearly two-dimensional flow at the trailing edge, characterized

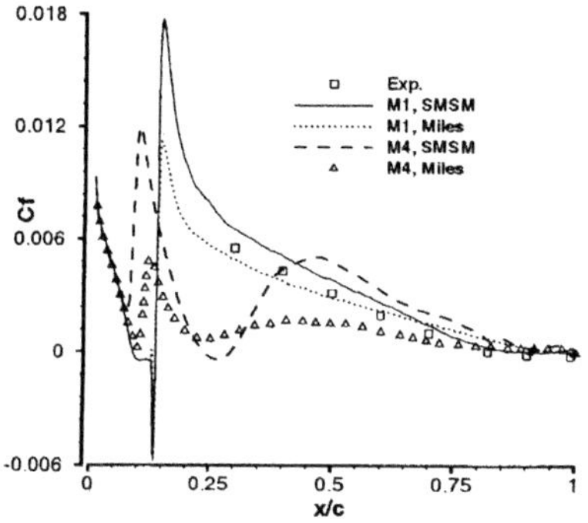

Figure 129 ONERA: Effect of SGS modelling and (x,y) mesh resolution on the mean skin friction coefficient C_f.

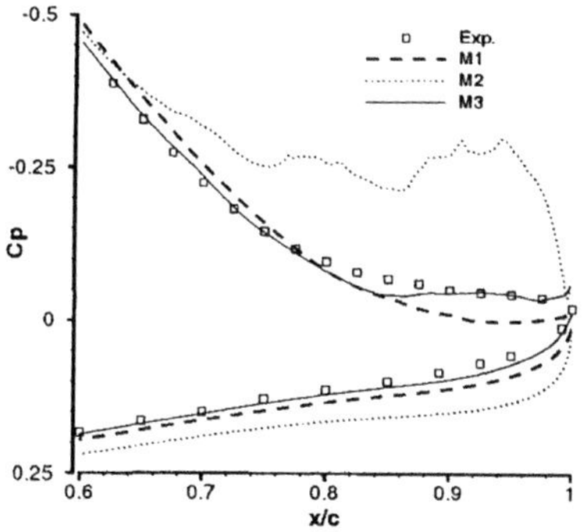

Figure 130 ONERA: Mean pressure coefficient C_p around the trailing edge using SMSM model.

by the presence of spurious intense vortices, which scale with the boundary layer thickness. As mentioned in Ref.[91], the large time scale of these eddies decreases the rate of statistical convergence, leading to the small irregularities observed in the suction side C_p-distribution.

Fig. 131 shows the mean C_f-distribution on the suction side. It appears that the separation bubble, located between $x/c = 0.09$ and $x/c = 0.13$, is not affected by the use of the different meshes. Moreover the characteristic features of a laminar separation bubble described by Alam and Sandham [92] are well recovered: there is a flat distribution up to $x/c = 0.12$ corresponding to the dead-air region and a large negative peak associated to the reverse flow vortex. This good physical behaviour can be related to the streamwise mesh resolution, since around 110 mesh points are located in the separation zone, followed by 40 points up to the positive peak arising due to the turbulent reattachment at $x/c = 0.15$.

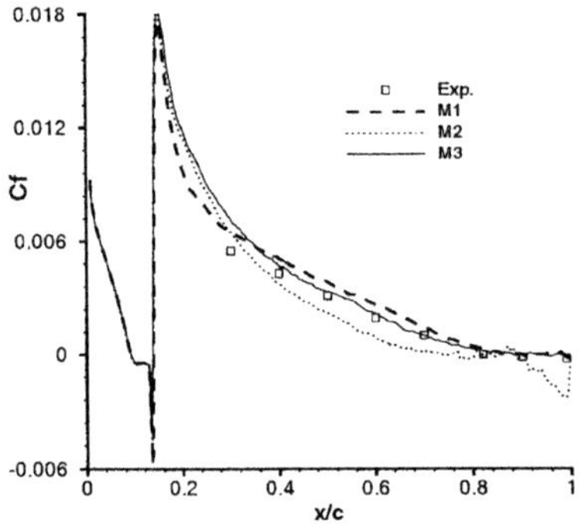

Figure 131 ONERA: Mean skin friction coefficient C_f on the suction side of the airfoil using SMSM model.

Downstream of this position the solutions for M1, M2, M3 start to differ. As for the C_p-distribution, the M3 simulation gives a very good agreement with experimental skin friction measurements. Indeed the largest error, located at $x/c = 0.3$, is less than 28%. The experimental value at this position is not very accurate because the experimental evaluation of the skin friction relies on the hypothesis of the existence of a logarithmic layer, whereas the boundary layer is still in non-equilibrium [90]. Therefore the observable error is less than 6% for the most reliable measurements (between $x/c = 0.5$ and $x/c = 0.7$). The use of a reduced spanwise extent in the M2 simulation does not modify the friction coefficient up to $x/c = 0.25$. But after this point, the spanwise blockage of the turbulent structures implies an undershoot of C_f, yielding a too vigorous reverse flow at the trailing edge. The use of a coarser spanwise resolution in the M1 simulation modifies the skin friction coefficient in a more complex way. Two zones could be distinguished: up to $x/c = 0.35$ the insufficient resolution does not allow the accurate representation of the turbulent structures, leading to an undershoot of C_f; further downstream C_f is too large, but the slope of the M3 results is almost recovered, which may indicate that the resolution is sufficient in this zone.

Figs. 132 and 133 compare the computational and the experimental (where available) mean streamwise velocity profiles in the wall-normal coordinate system located between $x/c = 0.1$ and $x/c = 0.99$. It appears that up to $x/c = 0.2$ the profiles are not sensitive to the use of the different meshes. Once again, the M3 simulation is in very good agreement with experimental profiles since the growth and shape of the boundary layer are well recovered at all locations. The use of the coarser spanwise resolution in the M1 simulation leads to fuller profiles in the near-wall region, which delays the separation in the trailing edge zone. On the contrary, larger shape factors are obtained with the shorter spanwise extent in the M2 simulation, leading to a too early separation around $x/c = 0.75$.

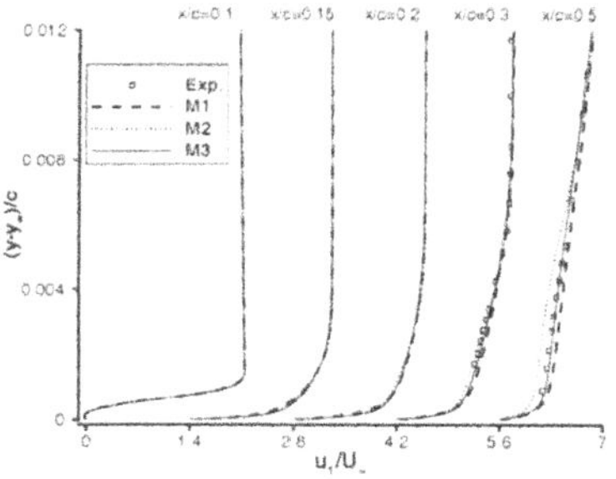

Figure 132 ONERA: Mean streamwise velocity profiles obtained with the SMSM in terms of the normalized wall normal distance; individual profiles are separated by a horizontal offset of 1.4 with the corresponding zero lines located at 0, 1.4,...,7.

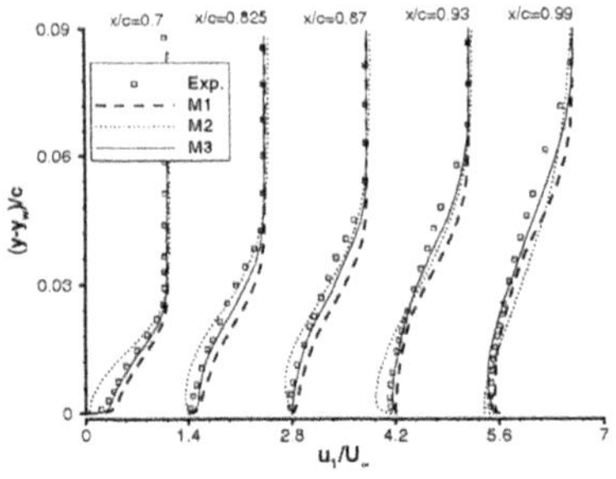

Figure 133 ONERA: Fig.132 continued

The root mean square (RMS) streamwise velocity profiles in the wall-normal coordinate system are plotted as a function of the normalized wall normal distance in Figs. 134-135, whereas the resolved shear stresses are presented in Figs. 136-137. For reasons of clarity the M2 results are not shown in Fig. 135 and Fig.137 at positions $x/c = 0.825-0.99$. Indeed, the too limited spanwise extent generates very intense non physical fluctuations typical of "two-dimensional" simulations [93], illustrated by the overestimation of up to 100% at $x/c = 0.7$. Further downstream these overshoots can even reach 250%, prohibiting a clear representation of all results in the same figure. From $x/c = 0.1$ to $x/c = 0.7$ the results of the different simulations are very close and in quite good agreement with the experimental data where for both the RMS streamwise velocity and resolved shear stress profiles.

The small fluctuations detected at $x/c = 0.1$ on Fig. 134, especially for the M2 simulation, indicate that the dead-air region in the laminar bubble oscillates slightly. Downstream of this position the intense fluctuations (close to 30% of u_∞) occurring at $x/c = 0.15$ in the turbulent reattachment zone relax slowly to a values typical of a turbulent boundary layer near equilibrium by $x/c = 0.5$.

The well known near-wall peaks are slightly larger with the M1 simulation but no experimental measurements are available so close to the wall, preventing the validation of this point. Even if the M1 and M3 results are still in quite good agreement with the experimental data at $x/c = 0.7$ significant discrepancies occur downstream in the separated region, especially on the M1 mesh for both the RMS streamwise velocity and resolved shear stresses profiles. This may be caused by both simulation and measurement errors. On the one hand, the spanwise extent of 0.012c becomes more and more ill-suited to the boundary layer thickness, which is close to $0.1c$ at $x/c = 0.99$. On

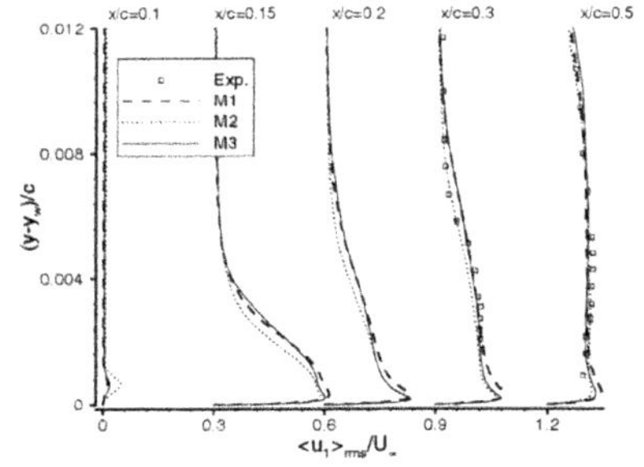

Figure 134 ONERA: Profile of the RMS streamwise velocity fluctuations obtained with the SMSM model; individual profiles are separated by horizontal offset of 0.3 with the corresponding zero lines located at 0, 0.3,...,1.2

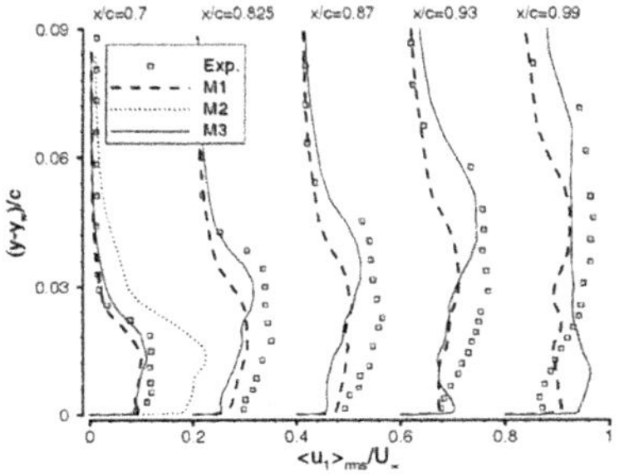

Figure 135 ONERA: Profile of the RMS streamwise velocity fluctuations obtained with the SMSM model; individual profiles are separated by horizontal offset of 0.2 with the corresponding zero lines located at 0, 0.2,...,1.

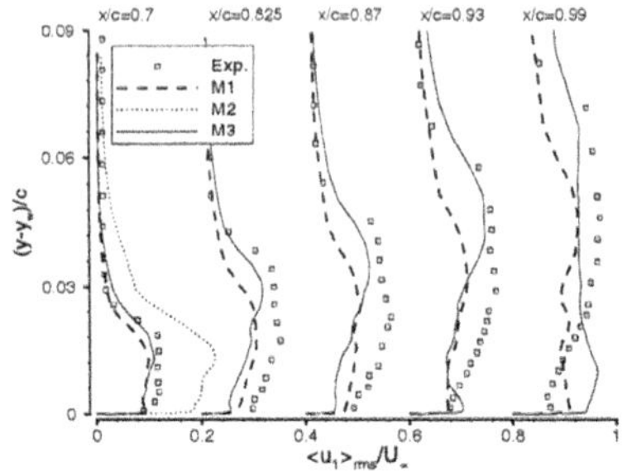

Figure 136 ONERA: Profile of the resolved shear stresses obtained with the SMSM model; individual profiles are separated by horizontal offset of 0.014 with the corresponding zero lines located at 0, 0.014,...,0.056

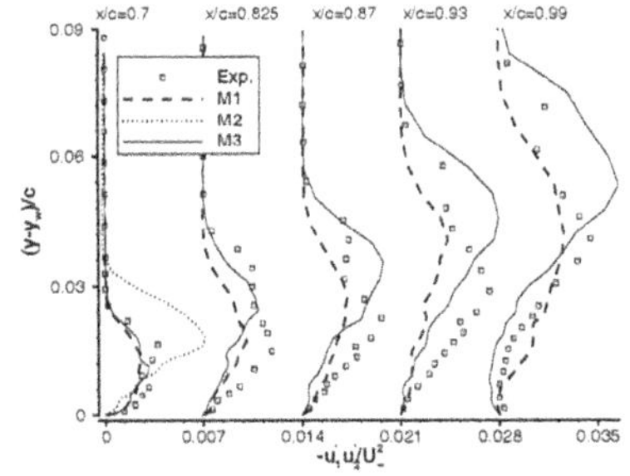

Figure 137 ONERA: Profile of the resolved shear stresses obtained with the SMSM model; individual profiles are separated by horizontal offset of 0.007 with the corresponding zero lines located at 0, 0.007,...,0.0028

the other hand, the accuracy of the RMS turbulence intensity measurements deteriorates when the local mean velocity has the same order of magnitude [90], as in the trailing edge separation zone. Another point is that in the windtunnel large-scale fluctuations due to the presence of sidewalls may be suspected.

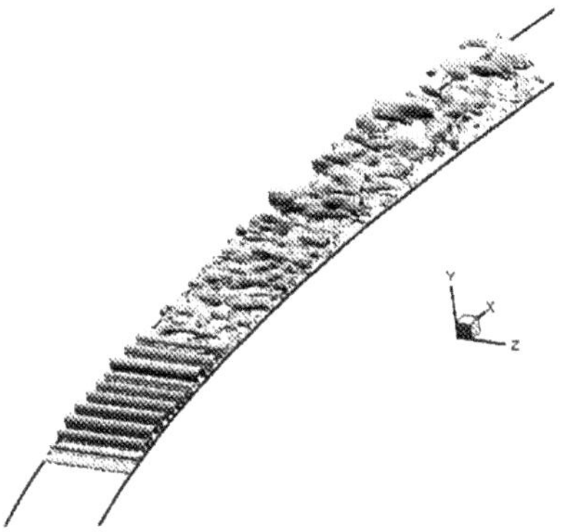

Figure 138 ONERA: 3D view of pressure fluctuation iso-surface in the transitional zone on the M1 mesh

Fig. 138 depicts a view of a pressure fluctuation iso-surface in the transitional zone of the M1 simulation. Contrary to the results of Weber and Ducros [56] which, due to insufficient resolution, were characterized by a "transitional" zone spreading over two or three cells, the present simulations on M1, M2, M3 recover a classical transition process: development of two-dimensional waves, followed by the formation of three-dimensional structures. As no artificial trip is used in the simulations numerical "noise" replaces the action of free-stream turbulence or surface roughness – the features which generate three-dimensional structures in practical applications.

The time history of the lift coefficient is shown in Fig. 139 for the period corresponding to the computation of the statistical moments. This allows to check that transient perturbations due to initial conditions are discarded for all simulations, since the signals are pseudo-periodic. Once again, the M2 simulation with the smaller spanwise extent strongly differs from the other simulations: if the mean lift coefficient is almost identical the RMS lift fluctuations are much larger. Table 4.8 summarizes computational and experimental results concerning the statistical properties of the lift and drag coefficients. Two sets of experimental data are presented corresponding to measurements undertaken in two different windtunnels. If the discrepancies associated with the lift coefficient is low, measurements of the drag coefficient differ by around 50%, emphasizing the poor reliability of these data.

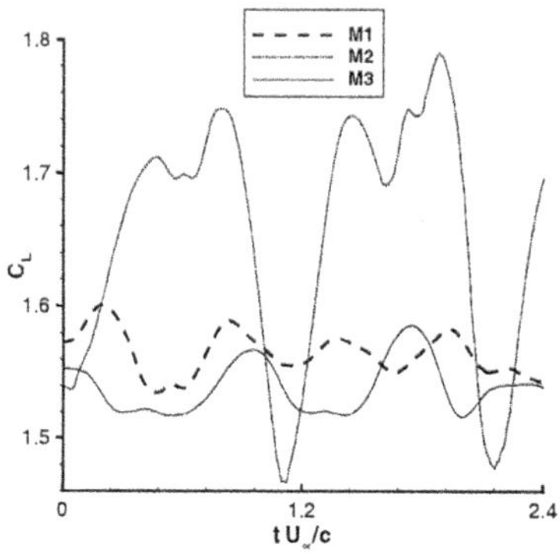

Figure 139 ONERA: Time history of the lift coefficient C_L obtained with the SMSM model. The time is normalized by c/U_∞

Quite good agreement is obtained for the lift coefficient on the M3 grid, whereas the drag is in the range of the two experimental values. As no separation occurs at the trailing edge for both the MiLES approach and the M4 grid the lift coefficient is overestimated in these simulations. Because of the too intense separation the M2 simulation gives a similar result. Concerning the fluctuations of the coefficients, the RMS value of the drag coefficients are much larger than those of the lift coefficient, as was observed by Weber and Ducros [56].

Conclusions. The influence of mesh resolution and SGS modelling have been assessed during the course of Task 5. For the same mesh resolution it appears from the different simulations that explicit SGS modelling performs better than the MiLES approach as long as the turbulent structures are sufficiently well described by the grid. But in case of a too coarse resolution the explicit SGS modelling performs similarly to the MiLES approach. The ef-

Table 30 ONERA: Lift and drag characteristics compared to experimental data.

	C_L	C_D	$< C_L >_{rms}/C_L$	$< C_D >_{rms}/C_D$
Exp. 1	1.548	0.0206		
Exp. 2	1.515	0.0308		
M1, SMSM	1.565	0.0262	0.007	0.191
M1, MiLES	1.602	0.0254	0.0037	0.118
M2, SMSM	1.653	0.0504	0.053	0.528
M3, SMSM	1.539	0.0250	0.0143	0.280
M4, SMSM	1.696	0.0210	0.009	0.143
M4, MiLES	1.725	0.0185	0.007	0.216

fects of both spanwise extent and resolution have been investigated, showing the substantial influence of these parameters. With the use of the largest extent and finest resolution the computed mean and fluctuating velocity profiles compared reasonably well with experimental measurements. To the authors' knowledge this constitutes the first satisfying LES of this complex flow.

4.9 Contribution by QMW

Authors: M.A. Leschziner, L. Temmerman

Overview. Building upon simulations undertaken earlier for the periodic-hill flow, QMW applied the methods developed in that context to the airfoil flow, subject to only minor changes. Specifically, no attempt was made to follow others in using a $2D/3D$ hybrid strategy, with the $2D$ portion covering the essentially inviscid far-field region. This decision was rooted in technical arguments noted in the conclusions of the overview chapter in this book relating to the airfoil simulations.

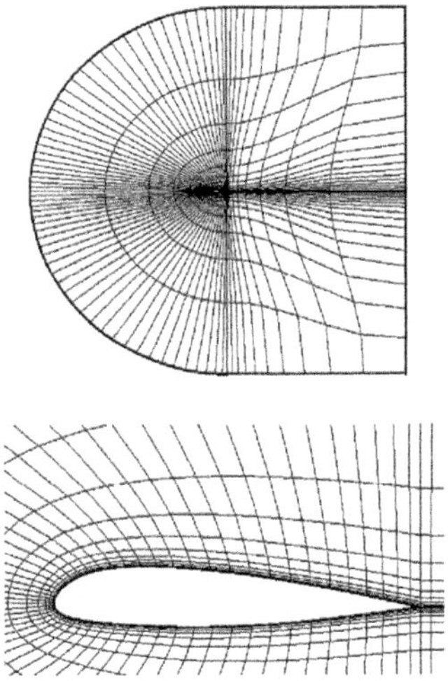

Figure 140 QMW: "UMIST mesh" (only every 3rd line included).

The majority of QMW's simulations were performed with variants of the basic "UMIST mesh" ($320 \times 64 \times 32$ nodes) shown in Figure 140, the variants being refinements that will be indicated later. This mesh was also adopted initially by some other LESFOIL partners, but was eventually replaced by a common mesh, a refinement of one proposed by ONERA. However, this latter mesh was found by QMW to give far slower convergence than the one proposed by UMIST. Hence, for the majority of the simulations, QMW preferred to continue using the UMIST mesh, which was eventually substantially refined in several stages, as will be explained below.

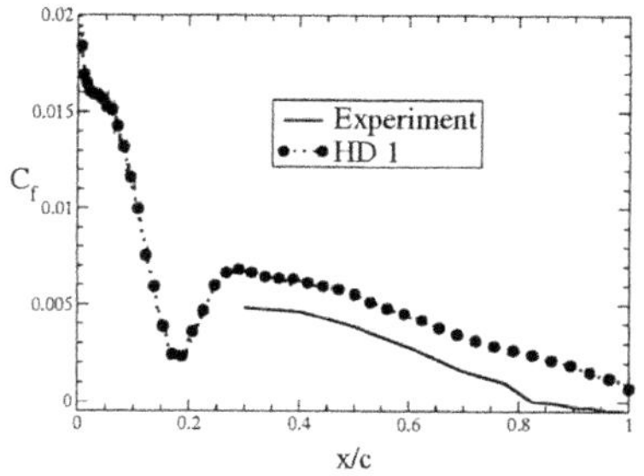

Figure 141 QMW: Distribution of suction-side friction coefficient for the densest grid (768 × 128 × 64).

Emphasis was placed by QMW primarily on investigations of the sensitivity of the simulation results to grid density and and the spanwise extent of the solution domain. Additional simulations were undertaken to clarify the sensitivity to the near-wall treatment and subgrid-scale modelling – although it must be said already here that these were hampered by resolution problems associated with grid coarseness. The finest grid used contained 6.4 million nodes.

In all simulations, a sharp trailing edge was specified, following a consensus within the LESFOIL Consortium that the introduction of the precise blunt edge did not modify by more than an insignificant margin the behaviour of the flow on the suction side. Moreover, most computations were done for a spanwise extent of 12% of chord, in line with a decision taken by the Consortium based on preliminary simulations. A spanwise extent well below this value was held to be too small to allow a proper resolution of the large $3D$ detached vortices following separation from the suction side.

In what follows, the simulation method is first summarised. The section after then presents selected results. Following a discussion of limitations associated with grid resolution and numerical practices, subsections focus specifically on the effect of the spanwise extent and the sensitivity to wall functions, conclusions are assembled in the final section.

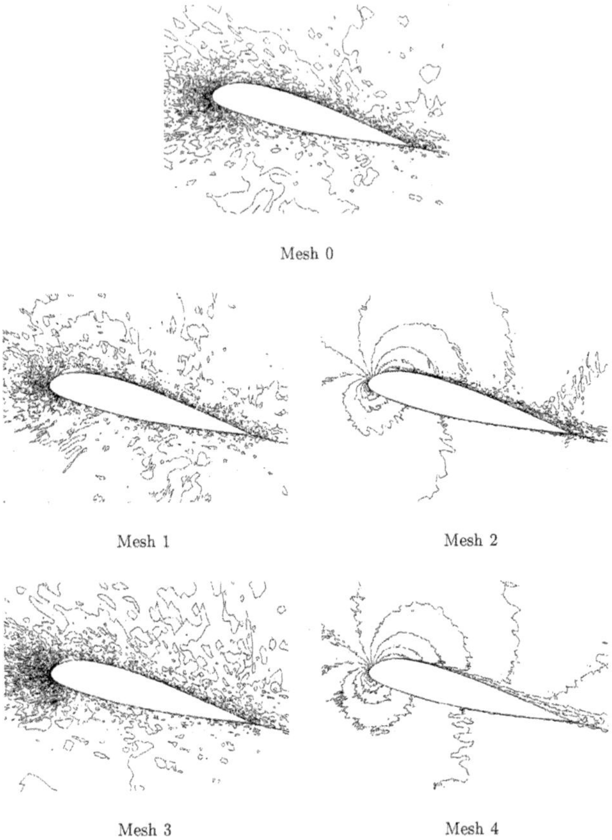

Figure 142 QMW: Instantaneous streamwise-velocity fields for Mesh 0 - Mesh 4.

The Numerical Method. The LES equations were solved using a second-order fractional-step method within a multiblock/multigrid non-orthogonal finite-volume approach, with the variables stored at cell centroids. The method first advances the convective and diffusive fluxes using the Adams-Bashforth scheme and then corrects the velocities, via the pressure, to satisfy zero-divergence. The time derivative is approximated by a second-order backward Euler scheme. The pressure is obtained by solving the pressure-Poisson equation using partial diagonalisation, which transforms the $3D$ problem into a set of $2D$ problems, each pertaining to one spanwise plane. In each spanwise plane, a V-cycle multigrid scheme and LSOR relaxation is used to solve the $2D$ pressure problem.

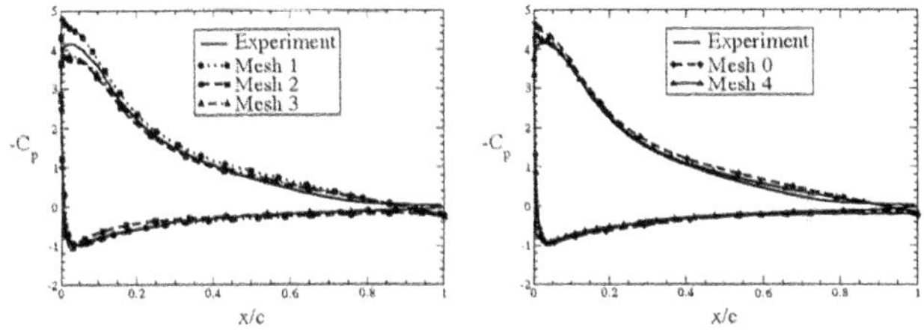

Figure 143 QMW: Distribution of the pressure coefficient for 5 different grid densities.

The code was fully parallelised. Most simulations were performed on a Cray T3E computer with up to 256 processors working in parallel. Typical parallel-efficiency figures were of order 90%. Details of the parallel implementation on various architectures and its effectiveness can be found in Temmerman *et al.* [94].

Uniform horizontal-flow conditions were prescribed on the outer grid line of the C-mesh. On the rear outlet boundary, non-reflective, convective outflow conditions were introduced, based on a formulation by Pauley *et al.* [95], a practice widely used in the LES community:

$$\frac{\partial u_i}{\partial t} + c\frac{\partial u_i}{\partial \xi} = 0 \tag{32}$$

where $i = 1, 2, 3$, c represents a convection velocity, usually taken equal to the mean flow velocity or the local flow velocity, and ξ is the direction normal to the boundary.

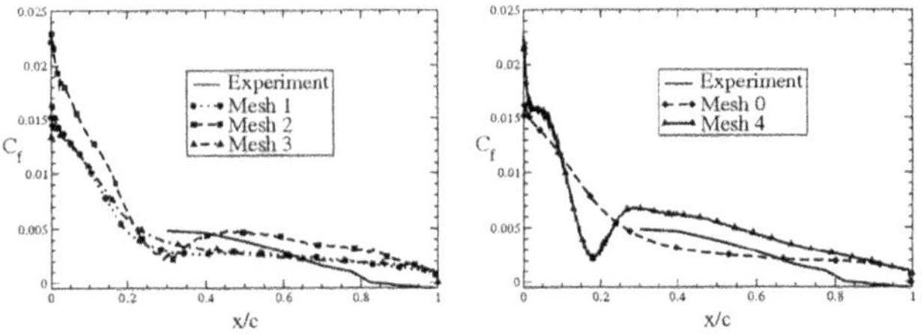

Figure 144 QMW: Distribution of the friction coefficient for the 5 meshes examined.

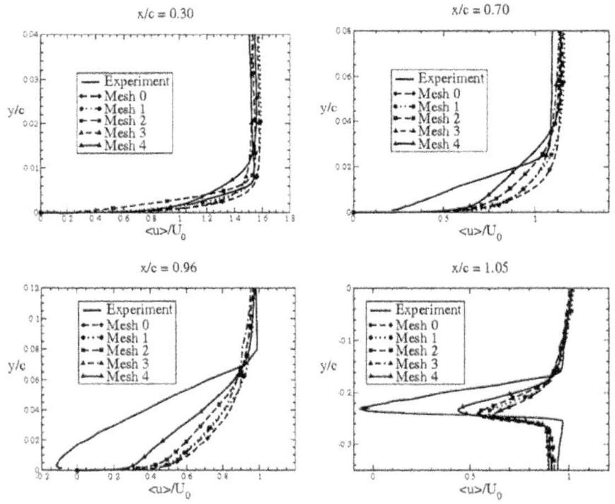

Figure 145 QMW: Profiles of mean streamwise velocity at 4 different locations for the 5 meshes examined.

Simulations were started from a potential-flow solution and were run for a period equivalent to 10 chord lengths, followed usually by 5 extra chord lengths over which statistics were assembled.

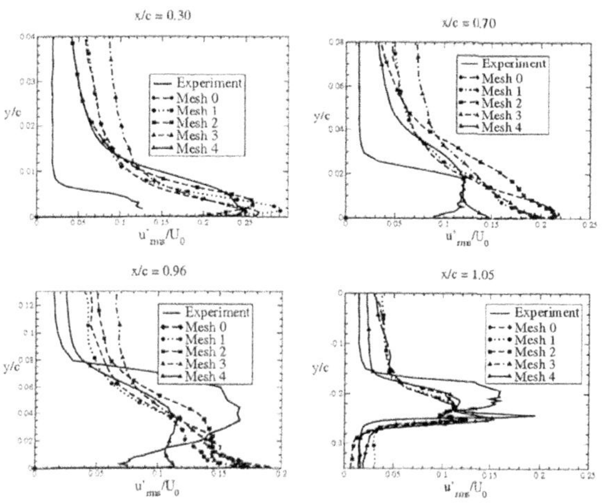

Figure 146 QMW: Profiles of r.m.s. of streamwise velocity at 4 different locations for the 5 meshes examined.

As will be demonstrated later, the exclusive use of central differences for convection provokes substantial numerical oscillations, especially in the irrotational region outside the boundary layers. In an effort to exercise control over these oscillations, central differencing in the sheared near-wall regions was blended with second-order upwind differences in the outer irrotational region. Details are given below. Suffice it to say here that the approach did not turn out to be satisfactory from a practical point of view. Apart from the difficulty of pre-defining the domain in which oscillation should evidently be damped, there is the more fundamental problem that central differencing is also likely to provoke oscillations in the rotational region at high Reynolds numbers. However, these may well remain undetected because of the presence of turbulence. Hence, the application of artificial damping based merely on geometrically pre-defined limits may be misleadingly 'effective' and will do nothing to remove 'numerical turbulence' in the shear layer, which is potentially much more damaging than oscillations in the outer irrotational region.

Resolution and Associated Limitations. Subsections to follow provide a detailed view of some specific parametric studies undertaken by QMW. The overall purpose of all simulations performed was to establish (or at least gain a view of) the sensitivity to resolution (grid density), the influence of varying the spanwise extent, the numerical benefits arising from introducing upwind-weighting in parts of the domain in which serious numerical oscillations arose with central differencing, and the response to using different subgrid-scale models and wall-law approximations.

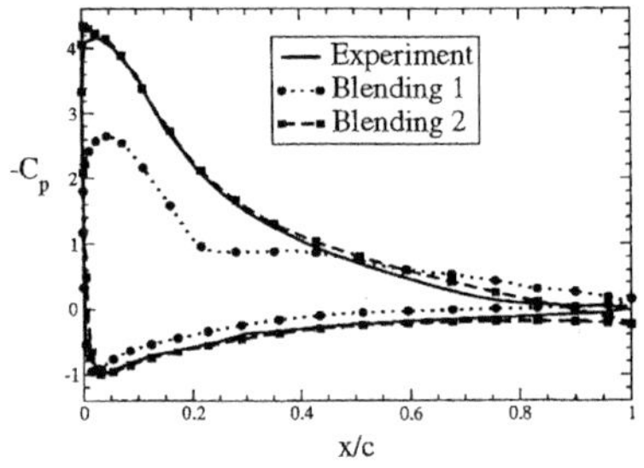

Figure 147 QMW: Distribution of the pressure coefficient obtained with two different blending practices.

The large majority of simulations were unfortunately found to be marred by two influential limitations, both simply arising from resolution constraints. First, transition is a crucially important process to capture or fix, for otherwise the predicted downstream behaviour is likely to be unrealistic. Experience by the writers and other partners indicated that extremely high resolution was required to actually capture transition. Clear evidence of transition was detected by the writers only when they used a grid of $768 \times 128 \times 64$ (6.4 million) nodes, covering 12% of chord. This is shown in Figure 141 by way of the related friction-factor distribution. Other flow properties computed with this grid are contained in figures to follow and in the overview chapter in this book. No attempt was made to fix transition in any of the computations. This could only have been done (and was indeed done by other partners) by suppressing turbulence by means of numerical dissipation up to a prescribed location downstream of the leading edge, a practice which is routinely adopted in RANS computation and has often been highlighted as a key limitation of turbulence modelling.

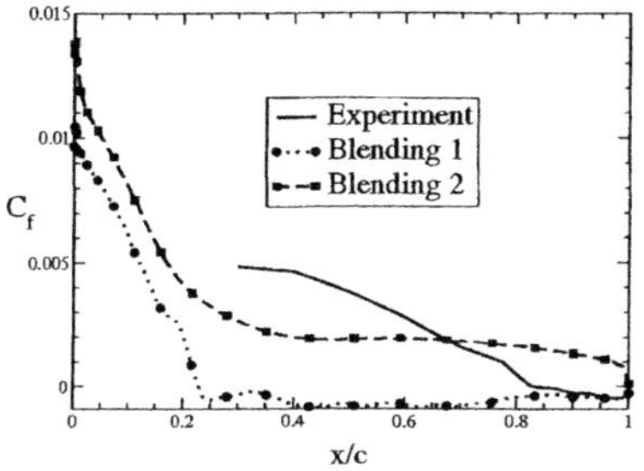

Figure 148 QMW: Distribution of the friction coefficient obtained with two different blending functions.

Second, as already noted above, the use of central differencing over the entire solution domain was found to give rise to numerical oscillations, which manifested themselves, among others, in pseudo turbulent normal stresses (but not shear stress!) outside the actual turbulent boundary layer. The severity of the oscillations was found to depend greatly on the grid density, with the normal-to-surface density being especially influential. This is illustrated in Figure 142 which gives snapshops of (instantaneous) streamwise velocity on the following 5 grids investigated:

- Mesh 0: 320 × 64 × 32 with 40 cells in the wake ("UMIST mesh");
- Mesh 1: 576 × 64 × 32 with 72 cells in the wake;
- Mesh 2: 320 × 128 × 32 with 40 cells in the wake, this mesh being obtained by dividing each normal-to-the-wall grid distance of the original mesh by two;
- Mesh 3: 320 × 64 × 64 with 40 cells in the wake;
- Mesh 4: 768 × 128 × 64 with 96 cells in the wake.

Mesh 1 - Mesh 4 arose from a selective refinement of the original UMIST mesh (Mesh 0). The lowest level of oscillations arose with Mesh 2 and Mesh 4, both containing 128 wall-normal grid lines.

Results for pressure coefficient, skin-friction coefficient, velocity profiles and streamwise rms velocity profiles are given, respectively, in Figures 143 to 146, while Table 31 provides additional information on lift and drag in comparison with the experimental data from both windtunnels F1 and F2. All simulations were performed over a spanwise distance of 12% of chord, with the Smagorinsky model ($C_s = 0.1$) with wall-damping ($A^+ = 25$) and the Werner-Wengle wall-function. As seen from Figure 146, all five grids mimic 'numerical turbulence' (albeit uncorrelated) outside the boundary layer. The lowest level tends to be produced by the grids with highest wall-normal density, at least upstream of $x/c = 0.7$, and this is consistent with the behaviour displayed in Figure 142. The densest grid, Mesh 4, gives the lowest r.m.s. level outside the boundary layer throughout the flow domain. Moreover, only the meshes with the highest wall-normal resolution, 128 lines, show evidence of transition around $x/c = 0.2 - 0.3$.

Table 31 QMW: Predicted drag and lift coefficients for the 5 meshes examined.

	C_D	C_L
Exp. F1	0.0204	1.56
Exp. F2	0.0308	1.515
Mesh 0	0.0280	1.693
Mesh 1	0.0197	1.713
Mesh 2	0.0383	1.475
Mesh 3	0.0665	1.550
Mesh 4	0.0437	1.553

As is evident from the above figures, none of the simulations comes close to giving a satisfactory representation of either the mean flow or the turbulence field. In particular, Figure 145 shows that all simulations fail to predict separation, although there is a gradual approach towards the experimental behaviour as the grid is refined, especially in the wall-normal direction. The lift coefficient is rather insensitive to the state of the boundary layer, in contrast to the drag coefficient, the value of which depends primarily on the skin friction as well as on the structure of the turbulent layer on the suction side. There is no systematic relationship between the predicted value of this coefficient and the grid, and the fairly close agreement achieved with Mesh 0 and Mesh 1 (depending upon whether experiment F1 or F2 is used as the basis for comparison) is fortuitous.

The trends displayed by the various grids provide some indications that an appropriate resolution might be achieved with a grid of order $1200 \times 240 \times 120$ (ie around 35 million) nodes. The simulations also point to the impossibility of investigating here with any degree of confidence the effectiveness of alternative SGS models or near-wall treatment, except in terms of indicating sensitivity in a qualitative sense.

An attempt was made to control the numerical oscillations outside the boundary layer by blending central differencing in the near-wall region with second-order upwinding in the irrotational outer region. This practice was also investigated by other partners [96], [97].

Blending was effected through:

$$\phi = \beta \times \phi_{centered} + (1 - \beta) \times \phi_{upwind} \tag{33}$$

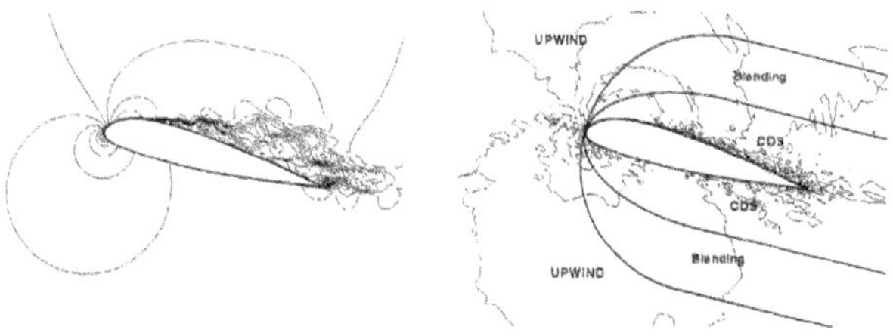

Blending 1 Blending 2

Figure 149 QMW: Instantaneous streamwise velocity field with two different blending strategies.

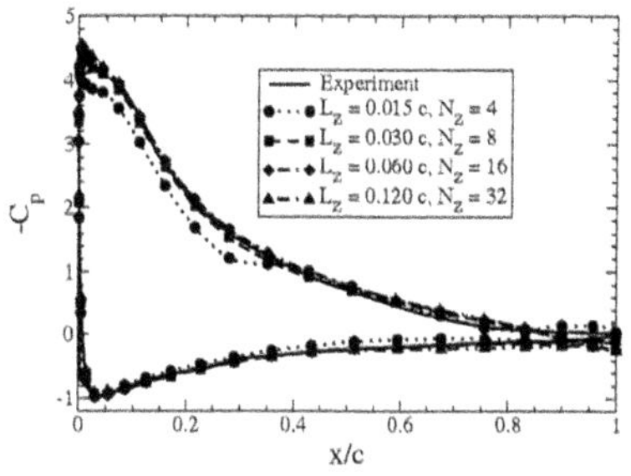

Figure 150 QMW: Distribution of the pressure coefficient for different spanwise slabs at invariant spanwise resolution.

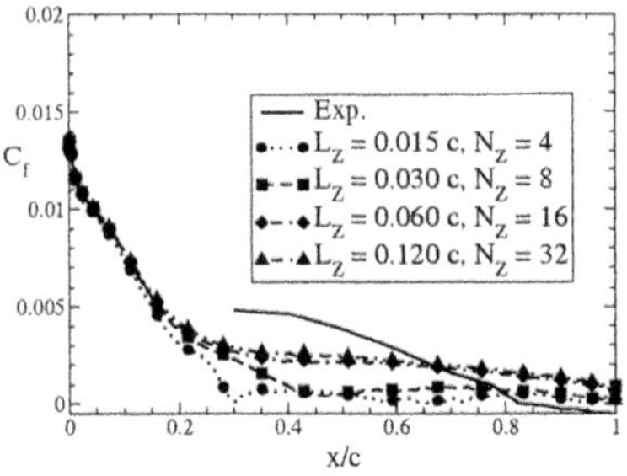

Figure 151 QMW: Distribution of the friction coefficient for different spanwise slabs at invariant spanwise resolution.

where β is chosen to lie between 0 and 1 and ϕ represents the convected variable. Evidently, $\beta = 1$ gives the fully centred scheme, while $\beta = 0$ gives the upwind scheme.

Two alternative practices were tried:

- Blending 1: the β coefficient value is controlled by the local Reynolds number ($Re_l = \frac{u\Delta x}{\nu}$):

$$\begin{cases} Re_l \leq 10 & \Longrightarrow \beta = 0 \\ 10 < Re_l \leq 30 & \Longrightarrow \beta = \frac{Re_l}{20} - 0.5 \\ Re_l > 30 & \Longrightarrow \beta = 1 \end{cases} \tag{34}$$

- Blending 2: β is only a function of the geometric coordinates, as indicated in Figure 149.

These two simulations were performed on the basic $320 \times 64 \times 32$ "UMIST mesh" with the Smagorinsky model ($C_s = 0.1$) and van Driest damping ($A^+ = 25$), and the Werner-Wengle wall treatment.

As illustrated by Figures 147 and 148, the two blending practices give very different results. Reynolds-number-based blending results in massive separation due to a failure to trigger turbulence – a behaviour similar to that observed by use of second-order upwinding throughout the domain. In contrast, geometric blending is considerably better in terms of predictive realism, but fails, as is evident from Figure 149, to eliminate the oscillatory behaviour in the outer irrotational region (although it alleviates it).

Overall, the introduction of artificial damping via upwinding cannot be regarded as a tenable practical approach, at least in the form attempted here. While it does have a stabilising effect on the computation, detrimental effects are difficult to avoid, and the precise details of the blending practice can strongly influence the physical realism of the solution. In addition, oscillations in the shear layer (from which damping is excluded) may be hidden by physical turbulence.

Mixing central and second-order upwind differencing, in a weight-ed-average sense, across the whole solution domain was found to be very effective, even with a 95%/5% relative weighting, but poses uncertainty in respect of numerical damping of turbulence in the shear layer. A more appropriate approach might be to use a "wiggle detector", such as that implemented by CERFACS. This would remove the arbitrariness of the present approach. However, a problem is here that the distinction between numerical wiggles and turbulent fluctuations is not clear. Whatever scheme is adopted, artificial dissipation is likely to remain a problem, and the only realistic solution is to substantially increase the grid density.

Effects of spanwise extent. As noted earlier, a spanwise distance of 12% of chord had been agreed as being the minimum necessary to capture the post-separation spanwise structure of flow with any degree of realism. Notwithstanding this agreement, most partners eventually opted for a lower distance, in one case as low as 1.2%. This gave rise to the important question as to whether the simulated flow changed drastically as the spanwise extent was reduced. QMW attempted to provide some pointers to the answer by performing simulations for spanwise distances of 1.5%, 3%, 6%, 12% of chord, keeping the spanwise resolution (in terms of z^+) invariant. However, a detrimental consequence of this invariant resolution was that the number of spanwise planes was reduced to 4 for the lowest spanwise slab. This clearly raises some serious doubts about the $3D$ nature of the simulation. It may be argued, however, that the observation of a major dependence of the flow features on the spanwise extent provides, even at that low resolution, an indication that this distance is too low.

As is demonstrated by the results in Figures 150 to 152, there appears to be a significant dependence on the spanwise distance below 6% of chord. This sensitivity must be viewed against the background of no separation being predicted. Had separation been resolved, the sensitivity might have been significantly more pronounced, as the scales of the influential structures rise rapidly with the onset of separation and the detachment of eddies. The simulations also indicate that the tendency towards separation grows as the spanwise extent is reduced to a level in which the simulation becomes almost two-dimensional in nature.

Effects of Wall-Treatment. The simulations performed for the hill flow (see Task 2 in Chapter II) demonstrated that the best results on coarse grids were achieved with the Werner-Wengle approach [98]. This wall treatment was therefore adopted in most of QMW's airfoil calculations as an alternative to

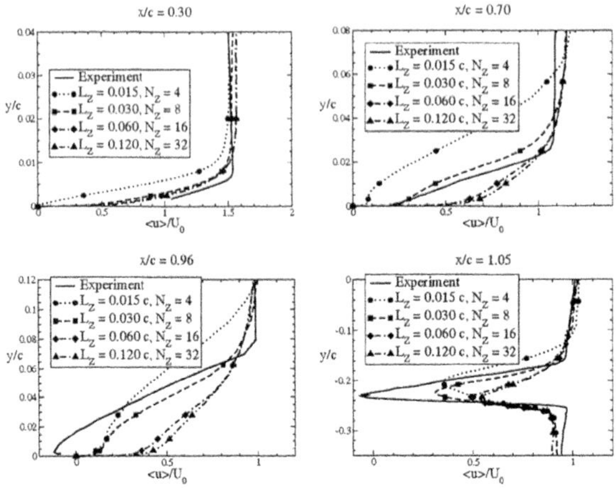

Figure 152 QMW: Averaged streamwise velocity at 4 different locations for different spanwise slabs at invariant spanwise resolution.

the three-layer log-law-based wall law and no-slip condition, the last shown to be a poor representation even at the much lower Reynolds number of the hill flow. Here, we report observations on the dependence of the simulations on the above three near-wall approximations.

The computations were done on the 320×64×32 mesh (Mesh 0) over a 12% spanwise slab with the Smagorinsky model ($C_s = 0.1$) and van Driest damping ($A^+ = 25$). It must be reiterated that the resolution limitations already highlighted prevent a reliable assessment of either the relative effectiveness of alternative near-wall approximations or of the influence of subgrid-scale models. Hence, the results reported below must be considered to convey, at best, qualitative messages only.

Table 32 QMW: Predictions of the averaged drag and lift coefficients with and without wall-treatment.

	C_D	C_L
Exp. F1	0.0204	1.56
Exp. F2	0.0308	1.515
No-slip	0.0404	1.682
Werner-Wengle	0.0301	1.682
3 Layers Log-Law	0.0405	1.627

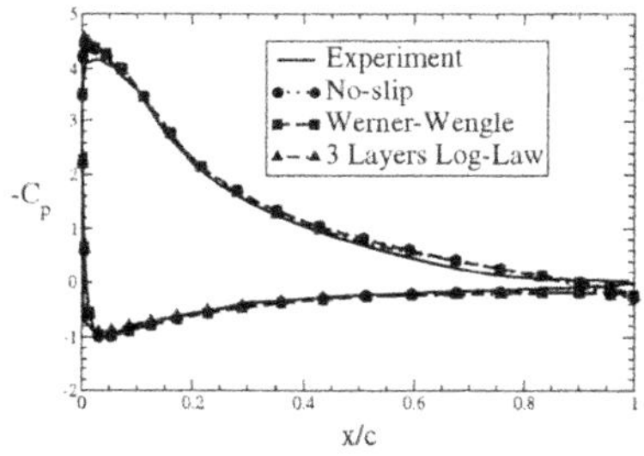

Figure 153 QMW: Distribution of the pressure coefficient with different wall-treatments.

Table 32 shows that the introduction of Werner-Wengle treatment appears to improves the prediction of the drag coefficient C_D, while the lift coefficient C_L remains too high. The pressure distribution around the airfoil, shown in Figure 153, is not greatly influenced. However, the friction coefficient in the first 40% of the airfoil is, as expected, quite sensitive to changes in the wall treatment, featuring a marked decrease upon the use of the Werner-Wengle approximation (see Figure 154).

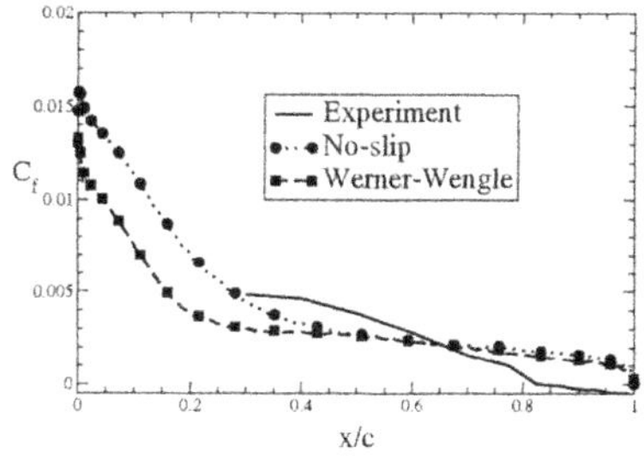

Figure 154 QMW: Distribution of the friction coefficient on the suction side with different wall-treatments.

Figure 155 shows that, except for the first upstream velocity profile, the introduction of the Werner-Wengle treatment leads to a slight improvement in the predictions, which may not, however, be quantitatively meaningful in view of the serious resolution limitations.

Overall, the introduction of the Werner-Wengle approximation may yield a marginally positive effect on the predictions, but it must be acknowledged that little can be achieved with, or reliable conclusions drawn about, alternative near-wall treatments, unless a much denser grid is used to improve resolution.

Conclusions Overall, the results achieved have been disappointing. The main obstacles are all due to insufficient resolution, including the presence of numerical oscillations. Although the introduction of artificial dissipation reduces the level of oscillations, the manner in which damping is introduced has been observed to exert a very strong influence on the predicted flow features, and the practice must, therefore, be viewed with considerable suspicion.

Despite the difficulties, it may be claimed that the results presented have shed some light on the influence of the various parameters which exert an influence on the quality of the results. It has been shown that a proper representation of the near-wall region by means of an adequate grid and/or by means of an adequate treatment of wall boundary condition is extremely important. The spanwise distance requires considerable care, as the study has provided indications of a strong sensitivity of the solution to this distance and the resolution within the spanwise slab. It must be acknowledged, however, that the investigation of the influence of the spanwise extent at a fixed resolution needs to be accompanied by additional studies in which the influence of increasing resolution for a given spanwise distance is quantified. Strictly, the former can only be safely identified if, for each spanwise distance, the spanwise resolution is sufficiently high to eliminate dependence on that resolution. Since, however, the spanwise grid density cannot be separated from the density in other directions, this is bound to be a difficult undertaking requiring many costly simulations.

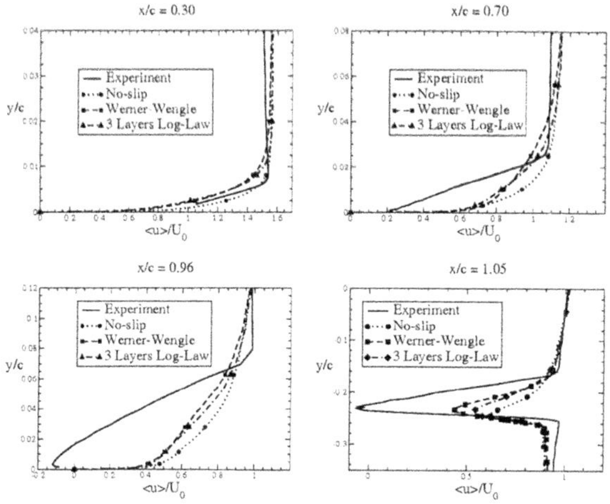

Figure 155 QMW: Averaged streamwise velocity profile at 4 different locations with different wall-treatments.

IV. LESSONS LEARNED

IV. LESSONS LEARNED

5 Synthesis of the Airfoil Flow Simulations

Authors: C.P. Mellen, J. Fröhlich, W. Rodi

The following synthesis covers the results obtained with LES and DES by CERFACS, Chalmers, Dassault, FLUENT, IfH at Karlsruhe, ONERA, and QMW/UMIST. Surrey could only perform very preliminary calculations which are not included here. The simulations performed by Alenia employed RANS models and will therefore be addressed in the final chapter of the book together with previous RANS calculations.

During the course of the project the partners tested many different combinations of parameters and models. The most important tests are described in their individual contributions. In the following we mainly concentrate on the most successful computations of each partner.

5.1 Common Mesh Comparisons

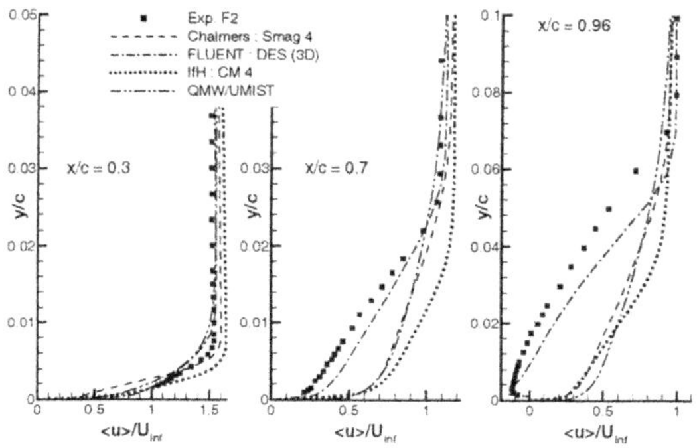

Figure 156 Common mesh, $< u > /U_\infty$ profiles

Fig.'s 156 to 161 display comparisons of the results supplied by partners who undertook common mesh computations. The FLUENT and QMW/UMIST results are from the coarser 'B' mesh, while Chalmers and the Karlsruhe used

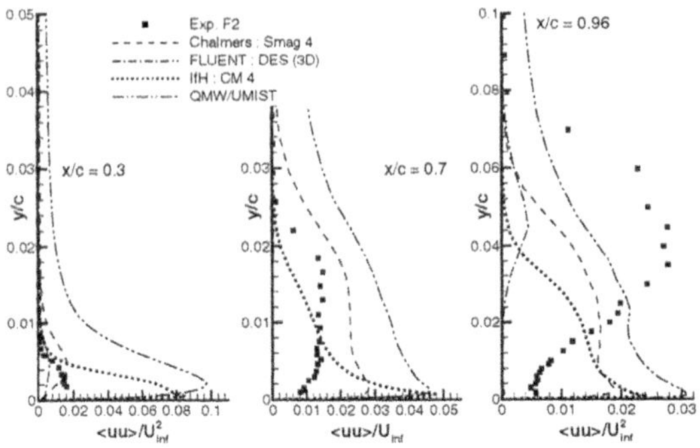

Figure 157 Common mesh, $< uu > /U_\infty^2$ profiles

the 'A' mesh. Details on these computations may be obtained from the individual contributions above.

Fig. 156 shows that only the FLUENT DES calculation was able to capture the trailing edge separation. The $< u > /U_\infty$ profiles from the other partners are much too full at both $x/c = 0.7$ and $x/c = 0.96$.

Fig.'s 157 to 159 show that generally the resolved stresses are also poorly captured. The FLUENT results are discussed later in the section on final results. The Chalmers, Karlsruhe and QMW/UMIST results show a distinctive near-wall peak in $< uu > /U_\infty^2$ that is not observed in the experiments. Such a peak is characteristic of an underresolved calculation. Interestingly, in the Chalmers profiles this peak is absent at $x/c = 0.3$ and, though it is present at $x/c = 0.7$ and $x/c = 0.96$, it is considerably lower than in the other partners' results. The reason is presumably the application of the Smagorinsky SGS model without near-wall damping of the turbulent viscosity ν_t in the Chalmers computations. This is not usual practice. For example, the QMW/UMIST computations, while also employing the Smagorinsky SGS model, use Van-Driest damping to enforce the correct near-wall asymptotic behaviour of ν_t. At $y^+ = 10$ the classical van Driest damping function equals 0.33 so that dropping this factor increases the turbulent viscosity at the first grid point by a factor of 3. At $y^+ = 20$ the value of the classical damping function is still 0.55. It can therefore be concluded that the underresolution, in particular at the first points adjacent to the wall, generates spurious streamwise oscillations which in the case of the Chalmers computations have been attenuated by additional turbulent viscosity close to the

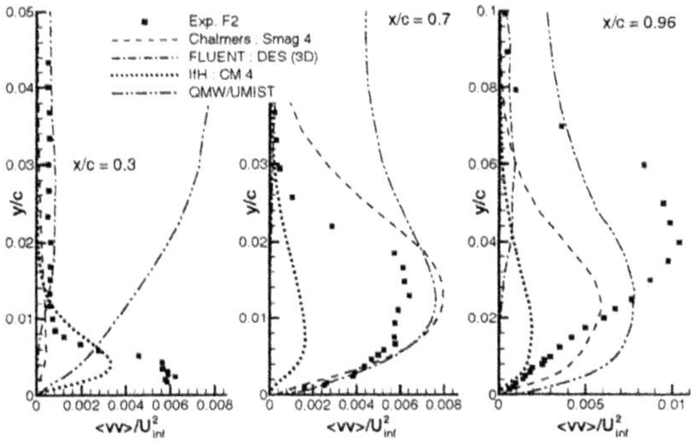

Figure 158 Common mesh, $< vv > / U_\infty$ profiles

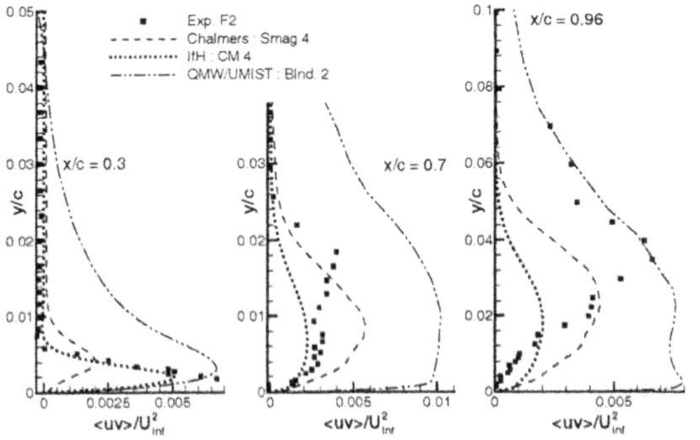

Figure 159 Common mesh, $< uv > / U_\infty^2$ profiles

wall. Comparing the Karlsruhe and Chalmers results shows that this effect renders the profiles more realistic (except those at $x/c = 0.3$ which are more influenced by the transition treatment).

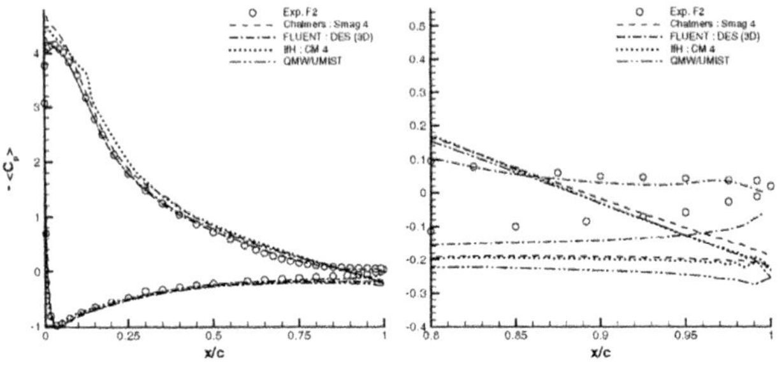

Figure 160 Common mesh. C_P vs. x/c (left); detail of C_P (right)

Fig.'s 160 illustrate the relationship between the trailing edge separation and the global pressure distribution. The FLUENT DES computation is by far the best match to the experimental data, being the only computation that indicates a pressure plateau at the trailing edge and does not overestimate the magnitude of the leading edge suction peak. Not surprisingly, the lift and drag coefficients returned by the DES (see Table 5.1) are also the most satisfactory.

Table 33 Common mesh, comparisons of C_L and C_D.

Partner	C_L	C_D
Chalmers Smag 4	1.70	0.0233
FLUENT DES–3D	1.568	0.0303
Karlsruhe CM 4	1.71	0.0158
QMW / UMIST	1.764	0.0429
Exp. F1	1.56	0.0204
Exp. F2	1.515	0.0308

Fig. 161 shows the distribution of the friction coefficient. Generally the Chalmers, Karlsruhe and QMW/UMIST results agree only poorly with the experimental data. In addition the Chalmers profile suggests that the computed boundary layer comes close to separating near $x/c = 0.35$. Downstream of $x/c \sim 0.4$ the FLUENT DES matches the experimental data reasonably well. Further, the FLUENT C_f distribution for locations past $x/c \sim 0.75$ illustrates the importance of capturing the trailing edge separation. However the lack of transition modelling in the FLUENT calculation and the coarseness of the common 'B' mesh is reflected in the their C_f distribution prior to $x/c \sim 0.2$.

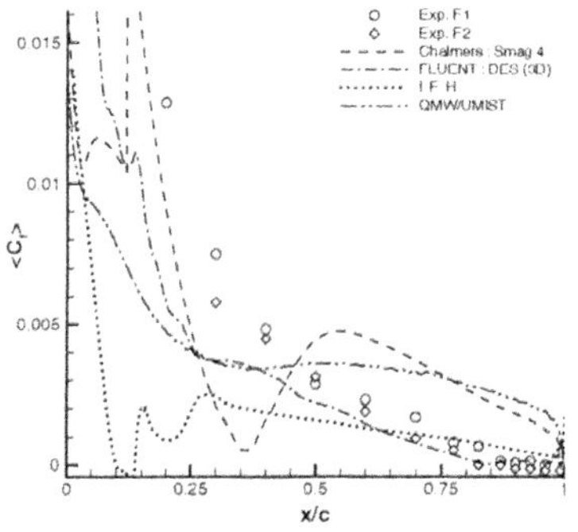

Figure 161 Common mesh, C_f vs. x/c

Initially, the aim of the common mesh computations was to enable a comparison of partner methodologies. Ultimately the common mesh computations could only serve to illustrate that LES is unlikely to provide satisfactory results when applied on meshes of inadequate spatial resolution.

5.2 Trailing edge geometry

The computations previously performed by RANS methods in [2] have all been performed with a sharp trailing edge. This makes grid generation much easier. The earlier grids used during the LESFOIL project such as the common grid employed for comparison in the previous section were also generated for a sharp trailing edge geometry. The experiment, on the other hand, was performed with a profile which had a blunt trailing edge of small but finite thickness ($0.005c$, see Fig. 162). This is fairly small compared to the thickness

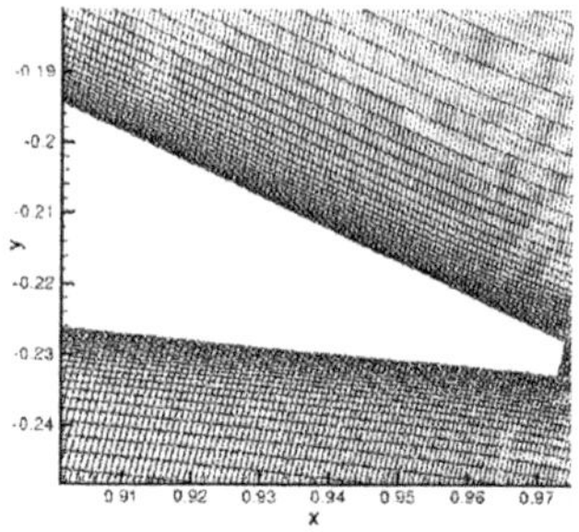

Figure 162 Trailing edge geometry of the A-profile used in the experiments and mesh employed in the Karlsruhe 2D3D computations. Strong zoom around the edge.

of the boundary layer at the trailing edge which is about $0.07c$. It is hence not expected that the presence of the blunt edge would influence the separation on the suction side close to the edge. On the other hand, the location of maximum backflow near the trailing edge in the experiment is about $0.003c$ remote from the profile (cf. Fig. 156). This is smaller than the thickness of the blunt edge so that an influence of this geometrical feature on the LES can not entirely be excluded.

To address this question, ONERA performed two computations under the same conditions for both the sharp and the blunt trailing edge geometry. In both cases the mesh was the one employed in their M1 simulation having the same distribution of points in the $x - y$ plane as the mesh M3 employed for the simulation reported here but a somewhat coarser spacing in z. The adaption of the grid to account for the blunt edge geometry was only very small. The results obtained are displayed in Figs. 163 and 164. They show that the presence of the blunt trailing edge does not alter to any significant extent the results of the LES. To avoid any concern about this issue many of the later meshes implemented the blunt edge as displayed in Table 34.

5.3 Final Results Comparisons

Table 34 compiles the parameters of those simulations considered to be each partner's most successful result. Details on these computations may be obtained from the contributions of individual partners. For FLUENT these data are identical to the ones displayed in the previous section as they have been obtained on the common mesh. Figures 165 to 170 report comparative plots of the results from the different partners.

As shown in Fig. 165, the ONERA M3 computation is the only one to return truly satisfactory velocity profiles over the entirety of the suction side

Table 34 Synopsis of parameters employed by the different partners for their best simulations. CM=common mesh, WFu=unstructured wall function mesh, *=unstructured cells in $x-y$-plane, bl=blunt, sh=sharp, WF=wall function, WN = white noise, NS= no slip, t.s.=transition strip, U =reference velocity, t_a = averaging time, (D)SM= (dynamic) Smagorinsky model, SA=Spalart-Allmaras model, SMSM=selective mixed scale model, WW=Werner Wengle wall function, WL=log-law wall function (2- or 3-layer). The grid spacings in wall units refer to maximum values after transition and are are meant to indicate the order of magnitude rather than precise values. n.p.m.=no particular measure taken, n.d.= no data available.

Partner, Run	mesh	T.edge	L_z/c	$N_x \times N_y \times N_z$	$\Delta_x^+ \Delta_y^+ \Delta_z^+$	wall
CERFACS Simu 2	ONERA	sh	0.02	$256 \times 64 \times 18$	1500 100 200	WW
Chalmers Smag 3	Chalmers	sh	0.08	$720 \times 65 \times 32$	550 50 350	WL
Dassault	CM WFu	bl	0.036	$8409^{*)} \times 25$	100	WL
FLUENT DES-3D	CM 'B'	sh	0.03	$360 \times 64 \times 16$	1100 10 300	DES
Karlsruhe 2D3D-3	Karlsruhe	bl	0.06	$2012 \times 72 \times 128$	100 10 40	WW
ONERA M3	ONERA	bl	0.012	$2760 \times 90 \times 72$	60 2 25	NS
QMW/UMIST M4	UMIST	sh	0.12	$768 \times 128 \times 64$	500 20 200	WW

Partner	Transition	Conv. Scheme	SGS	$\Delta t\, U/c$	CFL	$t_a U/c$
CERFACS Simu 2	trigg. by WN	Jameson +WD	WALE	$2.96\ 10^{-5}$	0.6	10
Chalmers Smag 3	blend. v.Leer	central	SM	$3\ 10^{-4}$	1	> 5
Dassault	n.p.m.	stabil. SUPG	sel. SM	n.d.	n.d.	6
FLUENT DES-3D	n.p.m.	central	SA-LES	$3\ 10^{-4}$	0.5	2
Karlsruhe 2D3D-3	blend. upw., t.s.	central	DSM	$1\ 10^{-4}$	0.65	> 10
ONERA M3 M3	resolved	AUSM+P + WD	SMSM	$3.3\ 10^{-5}$	2	2.4
QMW/ UMIST	n.p.m.	central	SM +v.D0.	$0.9\ 10^{-4}$	0.2	5

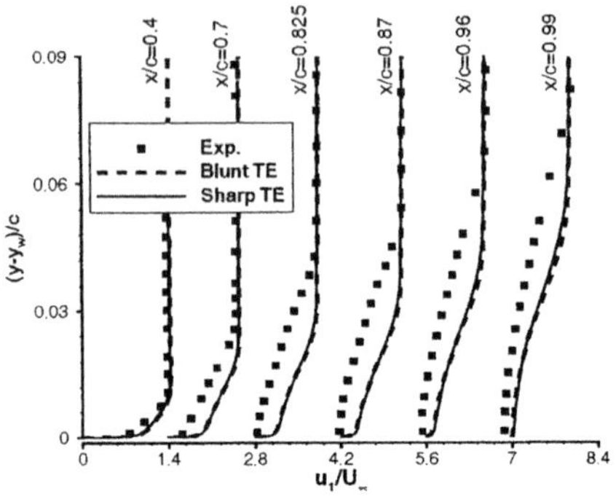

Figure 163 Mean streamwise velocity obtained with sharp and blunt trailing edge.

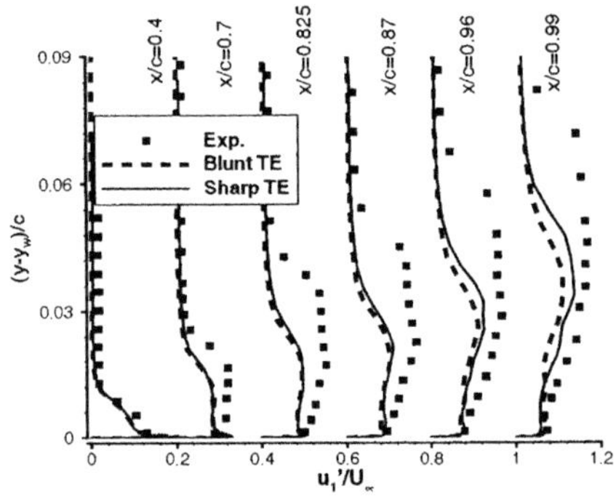

Figure 164 Streamwise velocity fluctuations obtained with sharp and blunt trailing edge.

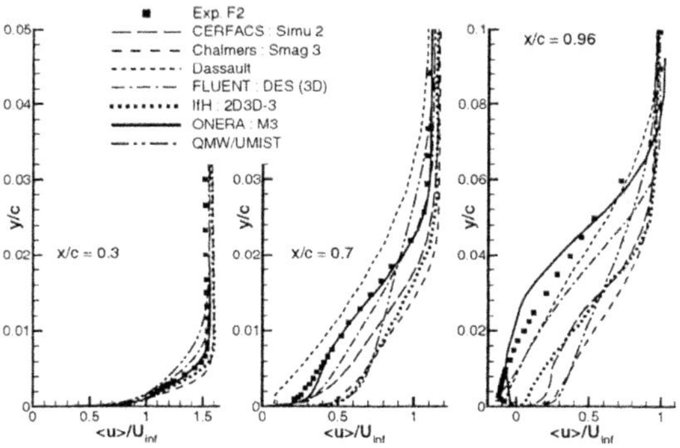

Figure 165 Final results, $< u > /U_\infty$ profiles

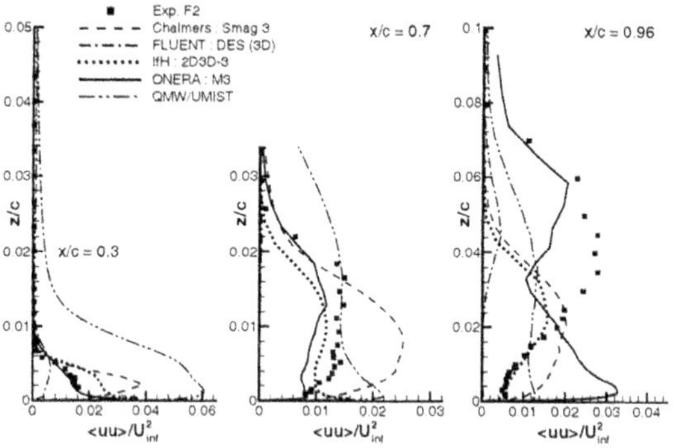

Figure 166 Final results, $< uu > /U_\infty^2$ profiles

surface. Of the remaining computations only those of FLUENT and Dassault indicate the presence of a trailing edge separation, though neither successfully capture the full magnitude of the reverse flow.

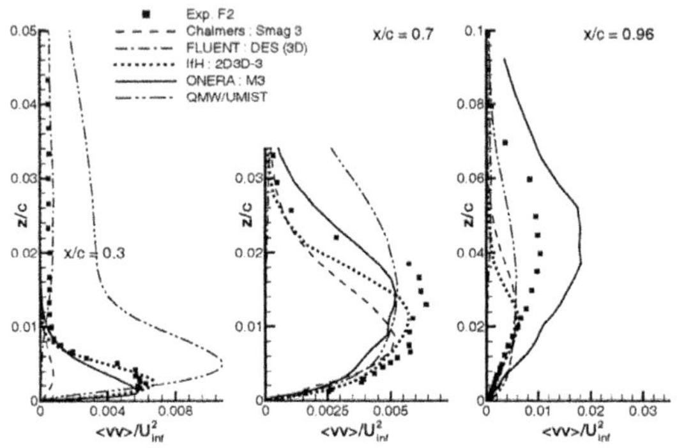

Figure 167 Final results, $< vv > /U_\infty$ profiles

All partners have difficulty in consistently predicting the fluctuating components $< u_i u_j > /U_\infty^2$ of the resolved velocity field. Prior to separation, at $x/c = 0.3$ and $x/c = 0.7$, the ONERA M3 computation is generally in best agreement with the experimental data (results are not available from all the partners and not always for all the three stresses). However, once separation occurs the limited spanwise extent of this computation ($L_z = 0.012c$ only) appears to become ill-suited to the thickening boundary layer. This imposes unphysical constraints on the large-scale structures and causes the resolved stresses to differ from the experimental data. Karlsruhe 2D3D-3 also predicts the fluctuations fairly realistically at $x/c = 0.3$ and 0.7 and near the wall also at $x/c = 0.96$. It is, however, noticeable that, as with most other partners, the streamwise fluctuations very near the wall are significantly overpredicted at the earlier stations. This is a characteristic which is often seen in LES computations that underresolve the flow. As is explained in Dassault's contribution, fluctuation results are not available for their computations with the selective Smagorinsky SGS model yielding their best mean-flow predictions included. In their contribution in Section 4.5, fluctuation results obtained with the standard Smagorinsky model are presented, and they are generally much larger than in the experiment. The authors blame insufficient mesh resolution for

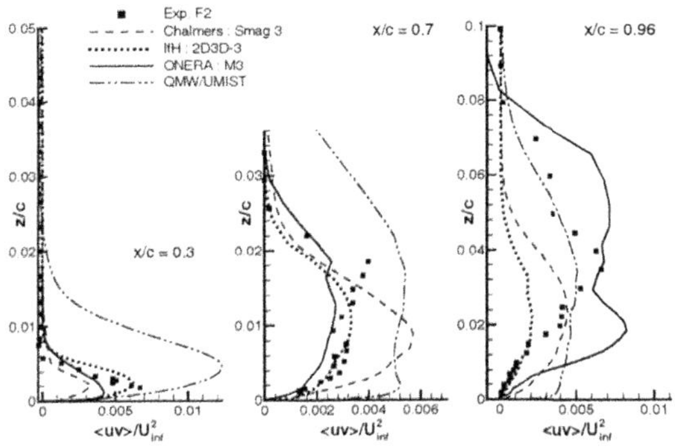

Figure 168 Final results, $< uv > /U_\infty^2$ profiles

this, and it seems unlikely that the fluctuations in the selective Smagorinsky model calculations are much smaller. In view of this it is somewhat surprising that the mean-flow field is reasonably well predicted. FLUENT DES on the other hand yields fluctuations $\langle uu \rangle$ and $\langle vv \rangle$ at $x/c = 0.7$ and 0.9 much smaller than the measured ones. It seems that here the resolved fluctuations do not contribute much to the momentum exchange and that it is mainly the model that contributes.

The FLUENT DES and ONERA computations best capture the global pressure distribution, see Fig.'s 169 and 170. This is consistent with a reasonably accurate representation of the trailing edge separation. It is noteworthy that the extreme resolution of the ONERA mesh has permitted to capture the small pressure plateau associated with the laminar separation that occurs at $x/c \sim 0.1$. This is a significant achievement, but at considerable expense.

The computations which do not yield the separation near the trailing edge also fail to capture the pressure plateau in this region and have a leading edge suction peak which is too large. An exception to this last observation is provided by the Karlsruhe 2D3D-3 result. Here the magnitude of the leading edge suction peak is noticeably smaller than that indicated experimentally. In the absence of a trailing edge separation the reasons for this result are not clear.

The pressure distribution from the Dassault computation indicates a trailing edge pressure plateau larger than that observed experimentally. Dassault attributes this to their inability to capture the laminar separation bubble in

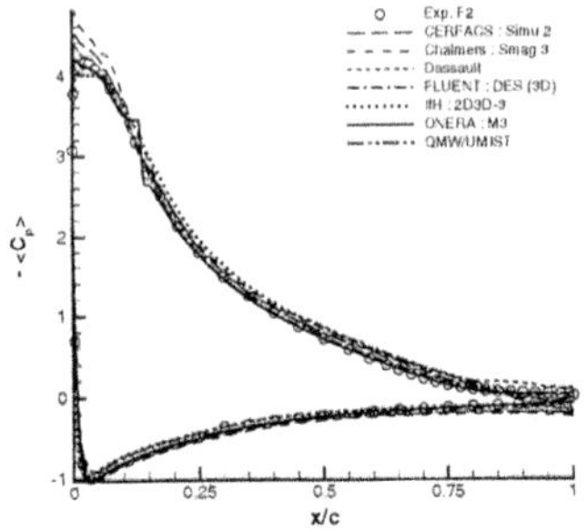

Figure 169 C_P vs. x/c

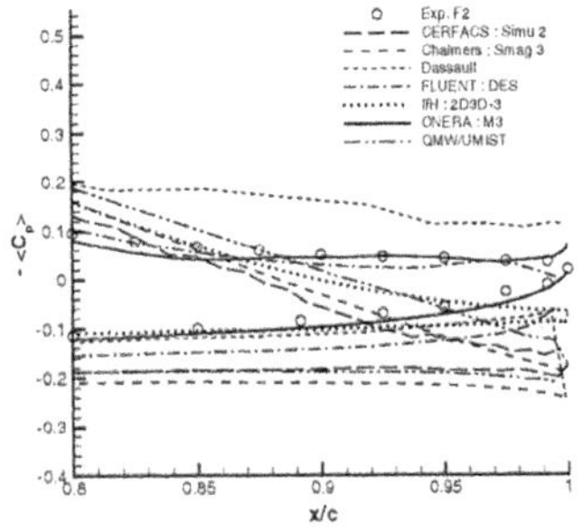

Figure 170 Detail of C_P vs. x/c

the transition region. The reasoning is that absence of this bubble alters the circulation and hence the pressure distribution.

Table 35 Final results, comparisons of C_L and C_D.

Partner	C_L	C_D
CERFACS Simu 2	1.655	0.0220
Chalmers Smag 3	1.72	0.0170
FLUENT DES–3D	1.568	0.0303
Karlsruhe 2D3D-3	1.54	0.0230
ONERA M3	1.539	0.0250
QMW / UMIST	1.682	0.0437
Exp. F1 [7]	1.56	0.0204
Exp. F2 [90]	1.515	0.0308

It is the ONERA M3 and FLUENT DES–3D computations which come closest to matching the experimental lift coefficient but Karlsruhe 2D3D-3 is also fairly close, (see Table 5.3). The ambiguity in the experimental drag data is unfortunate. FLUENT agrees very well with the F2 data, whereas ONERA, CERFACS and Karlsruhe 2D3D-3 take the middle ground.

The experimental skin friction data are matched extremely well by the results from the ONERA M3 computation, see Fig. 171. Noticeable is that the resolution of their mesh has enabled them to capture the reverse flow region of the leading edge laminar separation bubble. Of the other partners the Karlsruhe 2D3D-3 result achieves the best overall agreement with the experimental data prior to $x/c \sim 0.3$ where they also show the tendency to develop a laminar separation bubble. On the other hand, towards the trailing edge it is FLUENT that has the best match. It is clear to see that failure to predict the trailing edge separation leads to trailing edge C_f values that are much too large.

Concerning the spanwise extent L_z of the domain it is also worth observing that all the computations indicating separation at the trailing edge used comparatively small L_z. On one hand the limited spanwise extent does provide for increased spanwise resolution with the available computational resources as apparent from the numbers collected in Table 34. As discussed in Chapter 2 above, an adequate spanwise resolution is required to realistically represent the near-wall turbulent structures. On the other hand the small domain size in some computations raises questions about the spanwise correlation of these flow structures at the airfoil trailing edge and accordingly the amount to which the computed turbulence was affected by suppressing large three-dimensional structures. ONERA e.g. state that their result is influenced by this effect beyond $x/c = 0.8$. This is backed by the spanwise correlations determined in simulations of Chalmers with $L_z/c = 0.08$. At $x = 0.6c$ none

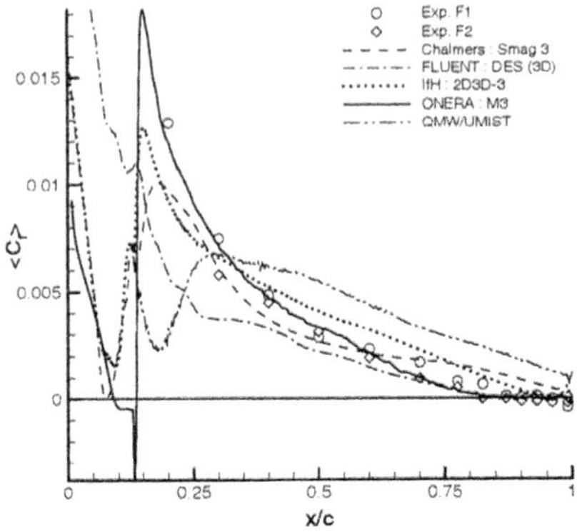

Figure 171 Final results, C_f vs. x/c

of the velocity correlations vanishes beyond $dz/c = 0.012$ and at $x = 0.9c$ zero crossings are located around $dz/c = 0.02$. The simulations by QMW with different spanwise extent and all other parameters conserved, even if performed on very coarse meshes, show a clear tendency towards separation when L_z is reduced. The extent to which the final results of the partners are influenced by this problem can however not be quantified further as they depend on many other parameters like grid resolution, grid stretching etc.

5.4 Subgrid-Scale Modelling

The subgrid-scale models used by the different partners during Task 5 are given in Table 5.4.

Quantitative analysis of the merits of each SGS model in the context of airfoil flows is made difficult by the variety of parameters – grid resolution, transition modelling, spanwise extent of the computational domain – which also influence the final result. Qualitative inferences may however be drawn.

On meshes which fail to adequately resolve the flow no SGS model seems able to return satisfactory results. Generally this is characterised by the inability to correctly predict the turbulent trailing edge separation indicated by the experiments and quite often by excessive velocity fluctuations in the near-wall region or unrealistic fluctuation levels. The most successful LES simulation, the ONERA M3 computation, also employed by far the finest mesh.

Table 36 SGS models employed in Task 5.

Partner	SGS Models
CERFACS	FSF, WALE, (DES)
Chalmers	Smagorinsky, Yoshizawa
Dassault	selective Smagorinsky
FLUENT	Spalart-Allmaras 1-eq. in DES
Karlsruhe	Smagorinsky, Dynamic Smag.
ONERA	SMSM, MILES (no model)
QMW	Smagorinsky

However, only one SGS model was evaluated on this mesh. In related computations these partners alternatively employed the MILES approach. This is based on the properties of certain numerical schemes to introduce numerical dissipation which can play a similar role as the dissipation introduced by the SGS model so that no explicit SGS model is employed. On a mesh coarser in spanwise direction than finally employed, ONERA found the use of an explicit SGS model better than the MILES approach. On coarse grids, the SGS model seems to have a strong impact on the solution, as illustrated by the contribution of Chalmers above, but no model yielded satisfactory results. On meshes which nearly fullfil the mesh resolution requirements of "classical" LES like the one of ONERA, the influence is weaker, but apparently the use of an accurate model can improve the predictions.

5.5 Near-Wall modelling

From the outset of the LESFOIL project, near-wall models were identified to be an important means of reducing the cost of the airfoil computations in Task 5. The models employed by each partner during this task are assembled in Table 5.5.

Table 37 Near-wall treatment employed by the partners in Task 5.

Partner	Near-Wall Treatment
CERFACS	No-Slip, Werner-Wengle, 2-layer log-law
Chalmers	No-Slip, 3-layer log-law
Dassault	log-law in "strong" form
FLUENT	DES
Karlsruhe	Werner-Wengle
ONERA	No-Slip
QMW	No-Slip, Werner-Wengle

Both the Werner-Wengle and 2- and 3-layer log-law near-wall models were originally formulated for use in attached boundary layers. Interest lay in observing the applicability of these functions to a boundary layer with a small separation region. Comparisons by partners of each of these near-wall models to simply using the no-slip boundary condition on nominal wall-function meshes indicate that both models generally improve predictions of quantities such as the skin friction coefficient C_f in the attached part of the boundary layer. Evaluation of their usefulness in the separated regions of the airfoil flow was however limited since in most of the partners' simulations the boundary layer actually remained attached.

This outcome should not necessarily be taken as an indication of the unsuitability of these approaches to regions of reverse flow as it is difficult to isolate the effect of any one parameter – such as SGS model, near-wall model or mesh resolution – on the outcome of a computation. There is little doubt, however, that the presence of the trailing edge separation does pose certain challenges to the useful application of either of these models. Karlsruhe argued that if a wall-function mesh is designed so that it is able to adequately resolve the region below the point of maximum backflow in the separation (a reasonable prerequisite) then y_1^+ of such a structured mesh must be so small (≤ 10 in the turbulent boundary layer) that both the Werner-Wengle model and the 2- and 3-layer log-law virtually reduce to a no-slip condition.

As mentioned previously a number of partners performed computations just using the no-slip boundary condition. Of these partners only ONERA and CERFACS (Simu 1) employed wall-normal spacing with the wall-adjacent grid point at a distance of $y^+ = O(1)$. Usually the use of a wall resolving mesh produces a marked increase in the cost of the computation due to the need to temporally resolve the fine near-wall fluctuations. This situation was reported by CERFACS and led them eventually discontinue their wall resolving computations. The CFL number needs to be about unity in order to temporally resolve all fluctuations that can be spatially resolved on the grid employed. ONERA used CFL numbers of about 2 which still fulfills this requirement. They furthermore report that some part of the computation has been repeated with an explicit scheme, which requires CFL of order unity, and obtained the same average results.

One reasonably successful wall-modelling calculation is the FLUENT DES. This approach employs a RANS turbulence model, here the Spalart-Allmaras (SA) ν_t-equation model, operated in unsteady mode, to represent the unresolved turbulent fluctuations close to the wall. In the interior of the flow field this model is converted into a one-equation SGS model by replacing the wall distance as length scale by the cell size of the grid. The computations by FLUENT have been performed on the common 'B' mesh with y^+ of the first grid point of order 10. In the spirit of DES the SA model would in fact require a wall-normal resolution of $y^+ = O(1)$. Also, the spanwise resolution seems very coarse (cf. Table 34), and it is not clear from the results to which extent the computed solution is actually three-dimensional and whether it is really an LES in the outer region or more a model-dominated simulation. The latter may explain why only the FLUENT DES calculation has been able to capture the trailing edge separation on the common grid as is apparent from Fig. 156. The $< u > /U_\infty$ profiles from the other partners in this figure are much too full at both $x/c = 0.7$ and $x/c = 0.96$.

As a result of the separation being obtained, lift and drag determined in the FLUENT simulation are also satisfactory. It is clear, however, that this computation could not be relied upon to give accurate information on the structure of the turbulence. The figures depicting slices at $x = const.$ in the individual report (Section 4.6) show that the mesh indeed is very coarse.

So while the FLUENT DES results for mean quantities are altogether better than the other LES computations on coarse grids, the experiences by other partners in the project do not back the superiority of the approach. CERFACS reports that their DES attempt employing the wall-resolving mesh generated in Task 1.1 was abortive as the computed flow remained very much steady-state in nature. Chalmers reported on unsatisfactory results of DES in the context of the hill testcase studied in Task 2 (wall modelling). Hence, further studies are needed to investigate the sensitivity of DES to parameters such as mesh resolution, time step and spanwise extent.

5.6 Transition Treatment

During the initial stages of the LESFOIL project it became clear that a proper simulation of the transition mechanisms by LES could not be aimed for in the project and that transition should be handled by special procedures with the aid of information from the experiments about the location of transition. During the course of the airfoil computations, however, it became readily apparent that transition handling was very closely linked to the issue of numerical oscillations appearing in nominally laminar regions of the flow. These unphysical oscillations arose due to a combination of the high-Reynolds-number flow and the use of centered differencing schemes on relatively coarse meshes. They were suppressed by using one of the following two approaches:

- Explicit application of an upwinded convection scheme.
- Use of a 'wiggle detector' which added numerical dissipation when an unphysical oscillation was detected. This detector is active in all regions of the flowfield.

The partners employing the first approach typically noted that once the calculations reverted to a centered convection scheme, the flow quickly became turbulent due to numerical noise arising from the discretization. This provided a means of *a priori* fixing the transition location. Usually a blending dependent on chord position was used to gradually switch between the convection schemes. Similar blendings were used to ensure further that subgrid-scale models and turbulent wall functions only contributed to the nominally turbulent regions of the flow.

When the second approach was employed, partners undertaking coarse-mesh computations found that transition had to be promoted through the introduction of numerical forcing at a point near the pre-specified transition location. This was done by CERFACS in some of their computations. ONERA found that on the very fine mesh used in their final calculations numerical forcing was not needed. Their resolution was in fact so good ($\Delta_y^+ \sim 2, \Delta_z^+ \sim 20, \Delta_x^+ \sim 100$) that they were able to simulate the actual transition processes – not in a DNS–like manner, but with some physical realism. In their calculations a thin laminar separation bubble developed between $x/c = 0.09$ and 0.13; transition occurred in the free shear layer of this bubble followed by turbulent reattachment, and all this agrees with observations in the experiments. Hence, in these calculations the location of transition was not fixed *a priori*, and this is in fact the only proper way to handle transition, but it needs a very fine resolution which could only be achieved by having a rather small spanwise extent of the calculation domain ($0.012c$).

The FLUENT DES calculations on a fairly coarse mesh did also not use an a priori specification of the transition location, but it is doubtful whether any transition actually occurred in these calculations. Judging from the C_f distribution, the straightforward application of the Spalart-Allmaras RANS model near the wall and the fairly poor near-wall resolution it is most likely that the boundary layer was turbulent right from the leading edge in these calculations. For going through transition with the Spalart-Allmaras RANS model, like with other RANS models, also some tripping is necessary [71].

In suppressing unphysical oscillations, the use of a wiggle detector seems slightly less *ad hoc* and more flexible than the explicit specification of regions in which an upwinded convection scheme should be applied. Possible concerns over such a detector arise due to the difficulty in quantifying just when and where numerical dissipation is added. Furthermore, this term in fact acts as an additional subgrid-scale term. Traditionally, however, LES practitioners prefer to rely solely on subgrid-scale modelling for the turbulent energy drain.

5.7 Synthesis conclusions

The Task 5 airfoil computations have brought significant advances in understanding the application of LES to airfoil flows at large Reynolds number and high angle of attack and of the problems associated with such calculations. It has become clear that an overall successful simulation of this type of flow can only be achieved with a well resolved LES which allows one to resolve the near-wall turbulent structures and return a proper simulation of transition. Results obtained with such calculations are largely satisfactory even though the spanwise extent of the calculation domain that could be afforded is really too small, especially in the downstream region with separated flow where larger structures develop. The finding that the trailing edge separation was only obtained in calculations with a relatively small spanwise extent, constricting the three-dimensionality of the large-small motions, is somewhat worrying. Unfortunately calculations using a wider spanwise slice of the airfoil would require even more computer resources than the very high ones just affordable in the calculations presented. However, novel techniques introduced by some partners, such as the blending of 2D and 3D calculations, will help to bring down the computer requirements.

The use of coarser grids in connection with wall functions ran into problems for the test case considered; in particular, the trailing-edge separation could not be reproduced in most of the calculations. The grid must still be fine enough to allow resolution of this separation region, but then the y^+-value at the first grid point is really too small to use wall functions while it is on the other hand too large for using the no-slip boundary condition. Also, in such calculations ad hoc methods must be used for handling transition which are not very satisfactory. Altogether it was found that the resolution has the greatest influence on the calculations and is the key to success while the subgrid-scale model appears to play a subordinate role.

Detached eddy simulations (DES) which have less stringent resolution requirements appear to have potential for calculating the airfoil flows in question; one such calculation was presented with promising results obtained for mean-flow quantities on a fairly coarse grid, but it appears the calculations were nearly two-dimensional and did not resolve most of the fluctuationing motion, also there was no transition; experiences with DES of other partners were not so successful. Due to this no final conclusions can be drawn and more tests are necessary.

V. CONCLUSIONS AND OUTLOOK

V. CONCLUSIONS AND OUTLOOK

LESFOIL set out to answer the question as to whether Large Eddy Simulation is, at present, a credible – or, indeed, superior – alternative to RANS modelling as a predictive approach to separated aerodynamic flows around streamlined bodies. It did so principally by reference to a particular single-element airfoil, with high incidence provoking a modest region of separation on the rear portion of the suction side. While this geometry is clearly very simple in terms of aeronautical practice, it is arguable that the focus adopted has not been overly restrictive, in so far as it has allowed important flow characteristics to be investigated which are encountered in aeronautical practice. The only major qualification relates to statistically three-dimensional features associated with spanwise mean motion which were absent in the present study.

LESFOIL was structured in a way that was planned to yield answers to specific questions about

- the role of mesh density and gradation;
- the effect of different numerical practices;
- the role of sub-grid-scale modelling;
- the role of resolution and modelling in the near-wall layer;
- the minimum size of the solution domain; and
- transition treatment

Precursor flows

A substantial proportion of the total effort was spent on precursor studies of plane channel flow and on the separated flow in a streamwise periodic channel segment with constrictions (hills) on one wall. While the relatively low Reynolds number of both flow groups allowed useful conclusions to be derived on subgrid-scale and near-wall modelling, the most important factor was, here too, the grid and its properties.

Wall-resolved simulations highlighted the need to capture the near-wall layer down to $y^+ = \mathcal{O}(2)$, with the grid density increasing toward the wall at a rate no bigger than 10% and with the aspect ratio being constrained to $\mathcal{O}(50)$. Moreover, the rate of change in streamwise grid density was found to be quite influential, and a ratio bigger than 10% was concluded to be inappropriate. The price of an increasingly liberal approach to these criteria was observed to be loss of accuracy, manifesting itself by a misrepresentation of the influential processes in the buffer layer, errors in the normal-stress anisotropy, the a wrong level of the log law in channel flow, due to errors in the wall-shear stress.

With the wall layer well resolved in plane channel flow, the subgrid-scale models performing best were dynamic variants and the WALE model, both types returning a broadly correct asymptotic near-wall decay of the subgrid-scale viscosity. This observation was influential in restricting the range of models investigated in the separated hill flow considered in subsequent efforts.

Initial channel-flow studies on the performance of wall-functions demonstrated that credible results for wall-shear stress and the inner-flow structure (in the log-law region) could be secured at a substantially lower resolution and thus saving in computing resource. Of the several log-law-based variants and the Werner-Wengle power-law-based approximation, the last was found to give the best results. However, the penalty with all formulations was a poor representation of the near-wall region, especially of the normal stresses. The fact that wall-law approximations returned broadly correct levels of wall-shear stress encouraged their adoption and further investigation in separated flow.

An attempt to achieve a superior performance by use of a hybrid RANS-LES strategy was only partially successful. A single practice was investigated, requiring the prescription of coupling conditions at the prescribed RANS/LES interface. Not surprisingly, these conditions were found to be highly influential, and the particular practices adopted gave rise to discontinuities or anomalies in the solution at the interface. Thus, the very limited work done in this area within LESFOIL must be regarded as inconclusive at this stage.

The investigations of the separated hill flow undoubtedly contributed greatly to a general appreciation of the challenges associated with resolving separation from curved surfaces. It is recalled that this flow was conceived as a benchmark for assessing a range of modelling practices that were expected to be unavoidable, as well as influential, at the high-Re airfoil flow. The decision to perform highly-resolved simulations to create a database for assessing coarser-grid results unfortunately necessitated a compromise in respect of the value of the Reynolds number. At the chosen value of 20500, and with the grid restricted, on resource grounds, to about 5M nodes, typical levels of the ratio of internodal distance to the Kolmogorov length was 5-10 over most of the flow domain. At this level of resolution, all important flow properties (dissipation rate being one exception) were, in effect, independent of subgrid-scale modelling. Thus, consistently, two independent simulations resulted in virtually identical solutions. The principal penalty of restricting the Reynolds number was the absence of a near-universal wall layer, preventing wall-laws from being tested reliably. In particular, the near-wall layer lacked a clear log-law region. In addition, the physical thickness of the near-wall layer over which wall-law approximations could be credibly tested (say, $y^+ = \mathcal{O}(30-50)$) was a significant proportion ($\mathcal{O}(5\%)$) of the thickness of the whole flow, making it difficult to prescribe a well-distributed grid with the wall-nearest gridline being at the requisite distance away from the wall.

Against this background, a range of investigations were undertaken to examine grid-density issues and sensitivity to subgrid-scale and near-wall modelling. One clear conclusion emerging from these studies was the overwhelming importance of adequate grid resolution for the accuracy of the simulations, especially (but by no means exclusively) in the region within which the flow separated from the hill. Importantly, streamwise resolution was observed to be highly influential, as well as the wall-normal distribution. Indeed, a clear correlation was observed between the predicted mean length of the recirculation zone and the streamwise resolution in the vicinity of the separation line. Although the Reynolds number of the flow was rather low, it was observed that a grid containing less than about 1M nodes afforded an insufficient resolution even for mean-flow quantities – although this conclusion was found to also depend, fairly sensitively, on the manner in which the near-wall layer was approximated. Importantly, this grid density pertained to simulations in which the near-wall layer was approximated by some approximate near-wall model, because such a grid did not allow full near-wall resolution. Thus, wall-resolved simulations would require a considerably higher grid density – though perhaps not the 5M nodes necessary to achieve de-facto freedom from subgrid-scale modelling.

Grid-resolution aside, the most influential issue in coarser-grid simulations of the hill flow was observed to be the near-wall approximation. Although the universal distance y^+ along the wall-nearest grid line was mostly within a value of 20 – with values greater than 12 confined to the region around the hill crest – the application of the no-slip condition was found to give seriously erroneous solutions, confirming the importance of representing realistically the buffer region. As noted already, the near-wall layer did not, unfortunately, lend itself to a well-founded investigation of wall laws. However, significant improvements in the results were observed upon the use of wall laws, instead of the no-slip condition. Of the variants investigated, the Werner-Wengle approximation performed best, perhaps because of its particular numerical implementation – specifically, the integration it involves in determining the relationship between the wall shear stress and the wall-layer-averaged velocity. This approximation was the only one giving coarse-grid results that were credibly close to the highly-resolved solution. A new near-wall model, based on back-flow scaling in the reverse-flow region, was also applied to the present flow. Although resting on fundamentally stronger grounds than conventional laws formulated for unidirectional flow, the performance of the back-flow model was not observed to be demonstrably superior to that of other variants. Similar comments apply in respect of the single hybrid RANS-LES methodology tested. This gave credible results, but not inherently superior to those obtained with wall-laws. The interface anomalies, noted above with respect to channel flow, were also observed in the separated hill flow, albeit diminished in severity.

Subgrid-scale modelling was found to be far less influential than the sensitivity to the near-wall approximation, and no clear trends could be identified, which would justify definitive recommendations. This is consistent with the observation that, at the Reynolds number in question, the subgrid-scale stresses constituted only a small proportion of the resolved stresses ($\mathcal{O}(2-5\%)$).

Airfoil flow

The principal effort within LESFOIL was that directed towards the Aerospatiale airfoil. This may be claimed to have demonstrated the major challenges LES faces in high-speed aerodynamic flows. These challenges are both numerical and physical, the former reflected by severe numerical oscillations with purely centred schemes that required the introduction of artificial damping, upwind-weighting and limiters. The use of this variety of numerical practices in the simulations performed by different partners and the evident sensitivity of the solution to them introduces, on its own, considerable uncertainty in respect of judging the role of modelling approximations.

Above everything else, the need for very high resolution, especially at the wall, emerged as being by far the most influential factor for predictive realism. In this respect, at least, there is virtually complete consistency with earlier observations derived from the hill-flow simulations. A major difficulty with the airfoil flow is the disparity in the scales of processes of similar importance to the overall flow behaviour. On the one hand, the laminar and transitional near-wall layer must be resolved, but this is extremely thin at the high Reynolds number in question. The resolution of transition is especially demanding, both in terms of wall normal and streamwise grid densities. On the other hand, the large structures associated with separation must be captured. Separation sets in gently and is thus highly sensitive to the evolution of the boundary layer as it approaches separation, the location of transition and the level of the circulation around the airfoil.

Credible wall-resolved simulations could only be realised with a numerical 2D/3D strategy in which only the near-airfoil region was resolved three-dimensionally. Apart from being only applicable to statistically 2D flow and requiring an a-priori decision on the 2D/3D domain partioning, the approach harbours some risks associated with the lack of resolution of 3D unsteady features induced in the inviscid region by the near-wall flow. However, this approach at least allowed sufficiently high grid densities to be achieved within unavoidable economic constraints, which suggest that the realism of the response of the boundary layer to adverse pressure gradient and its tendency towards separation improve monotonically as resolution improves. This conclusion is reasonably secure in relation to the attached boundary layer. Yet, following separation, the length scale of the influential structures increases rapidly, and the spanwise extent of the computational box used with the finest grids was too small to permit conclusions on the quality of the simulation in the separated region. Uncertainty in this respect was hightened by simulations suggesting a significant sensitivity of the separation process to the spanwise size of the simulation domain.

It has not been possible to derive clear conclusions on the role of subgrid-scale and near-wall modelling in the airfoil flow. Both appear to be influential, in so far as the simulated solution is materially sensitive to changing the models. However, no single combination could be identified as being superior to others. In view of the need to resolve the laminar layer and the importance of transition, modelling the flow with wall functions appears unpromising, and the poor results obtained with wall-functions did not come as a surprise. In particular, none of the wall-function simulations resolved the separated flow. However, it must be emphasized that the wall-function grids were relatively coarse. It is therefore not possible to claim without ambiguity that the failure to resolve separation was (principally) due to defects in the wall approximations.

A single simulation was performed with the increasingly popular Detached Eddy Simulation (DES) strategy. While that simulation was found to yield a good representation of the mean flow, the modelled component of the predicted flow was observed to be overwhelmingly dominant relative to the resolved (unsteady) component. Hence, this computation cannot be regarded a DES, but must be viewed as being a variant of a RANS computation.

It is finally of some interest to make a few comparative observations on the relative capabilities of LES, as have emerged from the present investigations, with those of the range of RANS models investigated by Alenia in the present project and within the EU project ECARP [2]. Alenia evaluated two turbulence models, one linear eddy-viscosity model and one explicit algebraic stress model. Neither of the models was able to predict the separation region. The observation that the linear model did not predict separation is well in line with what was found in the ECARP project. However, the studies in the ECARP project – some undertaken by contributors to LESFOIL – demonstrated that advanced turbulence models can give a faithful representation of the response of the suction-side boundary layer to the increasing angle of attack, including the condition of maximum lift and beyond. In particular, second-moment closure and non-linear eddy-viscosity models do relatively well, without ad-hoc corrections designed to limit the level of turbulence transport as separation is approached. However, there are many qualifications which may be fairly attached to this general conclusion. These are concerned with the generality of the models used and their calibration, the precise model variant used and the associated set of coefficients, the influence of 3D contamination (not examined within ECARP) and the treatment of transition (prescribed in all RANS studies). A serious obstacle to deriving reasonably unambiguous conclusions from the RANS studies is the evident high level of sensitivity of the predicted flow to minor model variations. Correspondingly, the experimental flow behaviour was also materially sensitive to minor variations in tunnel conditions and incidence angle. In some sense, the flow may be said to be pathological in nature, but it is also, unfortunately, important in a practical aeronautical context.

Based on the present study, it cannot be said that LES currently offers a secure means of diminishing the uncertainties associated with RANS modelling in mildly separated wing flows. Here too, the computed behaviour is very sensitive to a whole host of computational parameters. Most disconcertingly, the study suggests that the cost of obtaining a realistic representation of such flows (including transition) would be prohibitive in a practical environment: the ratio of LES-to-(2D)RANS resources, arising from the most realistic (ie demanding) computations for the A-airfoil in both categories, is of the order 10000. This ratio may not be lower in a fully 3D wing flow, despite the fact that the RANS computations would also need to be three-dimensional. This is because the hybrid 2D/3D strategy, as used here profitably for the present A-airfoil, would obviously no longer apply, and because the resolution requirements of LES in the spanwise direction are likely to be far higher than those for the corresponding RANS computations.

Thus, in summary, the single most important message is that the simulation of demanding aerodynamic flows featuring mild separation that is associated with progressive stall at high incidence requires extremely high resolution which is not, at this time, economically tenable for real fully 3D flows. The single most important issue is the near-wall region and its resolution with methods that are more economical than a wall-resolving simulation, yet offer a sufficient degree of physical realism.

Outlook and Recommendations

The overriding conclusion emerging from the study documented in this book is that adequate numerical resolution, especially at walls, is the key to LES yielding acceptable predictive realism in complex flow conditions in which viscous near-wall effects are important. It must be stressed that this is contrast to flows which are largely governed by processes that are remote from walls. Unfortunately, "adequate" translates to "economically untenable" at the high Reynolds numbers encountered in aeronautical practice, unless the near-wall region is bridged by some approximate treatment.

While it is often claimed that computer-resource limitations are a transient problem, which will eventually be solved by the seemingly constant rate at which computer speed increases and computer cost decreases, the reality is that industry is unlikely to be willing to tolerate the enormous cost and resource gap that will continue to separate RANS and properly resolved LES. There is also the problem that computer memory, essential for large-scale simulations, remains especially expensive and that there is a perceptible trend of HPC manufacturers towards producing high-speed machines with relatively low memory. This severely limits opportunities for highly-resolved simulations. Massive parallelism offers, in principle, access to high speed and large memory, but a systems with thousands of processors tend to be unwieldy and are affected by communication-bandwidth limitations. Industry is unlikely to acquire such systems.

The key question is then whether wall-resolution requirements can be substantially lowered by means of novel strategies without seriously detrimental effects on accuracy. This question is not easily answered on the basis of current knowledge. Wall functions have been extensively explored in the LES-

FOIL project, but are highly inaccurate in conditions far from equilibrium. A substantial amount of research is in progress on LES-RANS hybrids in which the near-wall region is computed by an unsteady RANS scheme. The basic rationale of this strategy is that a substantial removal of unsteadiness in the near wall region, with most of the turbulence activity delegated to a statistical model, would allow cells of much higher aspect ratios to be used than is possible in wall-resolved LES. However, the inevitable penalty is the loss of the turbulence structure in the RANS region and the unsteady dynamics which 'feeds' the regions resolved by LES. This is a tricky path in which an excessive level of near-wall damping of eddies may lead to a collapse of unsteadiness in the outer region. There are also a range of difficult issues to resolve in relation to coupling conditions at the interface and the location of the interface. Last but not least, there is no prospect of natural transition being resolved by a RANS scheme, and this implies the need for a zonal approach in which wall-resolved patches must be imbedded within a RANS-LES hybrid.

Against the background of current knowledge, research into hybrid LES-RANS strategies, with particular emphasis on near-wall modelling, must be strongly encouraged and supported, for it at least offers major savings in some flows without unacceptable loss of accuracy. What range of flows can be so computed is uncertain, as are the question of what level of modelling is required and whether heat transfer can be adequately predicted with hybrid schemes. Another research area which must be pursued is the development and use of high-accuracy unstructured- and adaptive-grid schemes which allow highly targeted near-wall refinement. However, while such capabilities offer significant savings over structured-grid schemes, the savings are not dramatic, because it is in the near-wall region that most of the resources need to be expended in wall-resolved simulations.

Over the past three decades, the demise of RANS and its progressive replacement with LES in industrial CFD have been proclaimed in numerous fora and on numerous occasions. These pronouncements are now recognised as having been overly ambitious. While major advances have undoubtedly been made in LES, it is becoming increasingly clear that LES is no panacea and faces some major challenges in difficult near-wall flows of particular interest in aerodynamics. It is only through sustained collaborative efforts of leading academic and industrial groups that these challenges are likely to be addressed effectively. Industry is likely to increasingly adopt LES for a range of problems in which mixing by large scales away from walls govern major operational characteristics, while RANS and/or possibly some forms of RANS-LES hybrids will be used for near-wall flows.

VI. REFERENCES

VI. REFERENCES

[1] Haase, W., F. Brandsma, E.E., Leschziner, M., Schwamborn, D., eds.: EUROVAL: A European Initiative on Validation of CFD-codes. Volume 42. Vieweg Verlag (1993)

[2] Haase, W., Chaput, E., Elsholz, E., Leschziner, M., Müller, U., eds.: ECARP – European Compuational Aerodynamics Research Projects: Validation of CFD-codes and Assessment of Turbulence Models. Volume 58. Vieweg Verlag (1997)

[3] Dervieux, A., Braza, M., Dussage, J.P., eds.: Computation and Comparison of Efficient Turbulence Models for Aeronautics – European Research Project ETMA. Volume 65. Vieweg Verlag (1998)

[4] Meier, H., Kreplin, H., Laundhausser, A., Baumgarten, D.: Mean velocity distributions in three-dimensional boundary layers developing on a 1:6 prolate spheroid with natural transition. report DFVLR IB 222-4/A 10, DFVLR, Göttingen (1984)

[5] Kreplin, H., Vollmers, H., Meier, H.: Wall shear stress measurements on an inclined prolate spheroid in the DFVLR 3mx3m low speed wind tunnel. report DFVLR IB 222-84/A 33, DFVLR, Göttingen (1985)

[6] Vollmers, H., Kreplin, H., , Meiera, H., Kühn, A.: Measured mean velocity field around a prolate spheroid at various cross sections. report DFVLR IB 221-85/A 08, DFVLR, Göttingen (1985)

[7] Huddeville, R., Piccin, O., Cassoudesalle, D.: Opération décrochage - mesurement de frottement sur profiles AS 239 et A 240 à la soufflerie F1 du CFM. Technical Report RT–OA 19/5025 (RT-DERAT 19/5025 DN), ONERA (1987)

[8] Kim, J., Moin, P., Moser, R.: Turbulence statistics in fully developed channel flow at low Reynolds number. Journal of Fluid Mechanics **177** (1987) 133–166

[9] 4th ERCOFTAC/IAHR/COST Workshop on Refined Flow Modelling. In: June 6-7, University of Karlsruhe, Karlsruhe. (1995)

[10] Almeida, M., Durau, D., Heitor, M.: Wake flows behind two-dimensional hills. Exp. Thermal and Fluid Science **7** (1993) 87

[11] Hanjalić, K., Obi, S., Hadžić, I.: 6th ERCOFTAC/IAHR/COST workshop on refined flow modelling: Wall-jets and flow over wall-mounted cubical obstacles. Report, ERCOFTAC Bullentin No. 36 (1998)

[12] Archambeau, F., Leschziner, M.: Vortex shedding past square cylinder. In: Proc. Workshop on Large Eddy Simulation of Flows past Bluff Bodies, Tegernsee, Germany (1995)

[13] Métais, O., Lesieur, M.: Spectral large-eddy simulation of isotropic and stably stratified turbulence. Journal of Fluid Mechanics **239** (1992) 157–194

[14] Piomelli, U., Liu, J.: Large-eddy simulation of rotating channel flow using a localized dynamic model. Physics of Fluids A **7** (1995) 839–848

[15] Davidson, L.: Large eddy simulation: A dynamic one-equation subgrid model for three-dimensional recirculating flow. In: 11th Int. Symp. on Turbulent Shear Flow. Volume 3., Grenoble (1997) 26.1–26.6

[16] Ducros, F., Nicoud, F., Poinsot, T.: Wall-adapting local eddy-viscosity models for simulations in complex geometries. In: 6th ICFD Conference on Numerical Methods for Fluid Dynamic. (1998) 293–299

[17] Sagaut, P.: Simulation numérique d'écoulements décollés avec des modèles de sous-maille. PhD thesis, Paris VI, Paris (1995)

[18] Werner, H., Wengle, H.: Large-eddy simulation of turbulent flow over and around a cube in a plane channel. In: Turbulent Shear Flows 8, Springer-Verlag (1991) 155–168

[19] Moser, R., Kim, J., Mansour, N.: Direct numerical simulation of turbulent channel flow up to $re_\tau = 590$. Physics of Fluids A **11** (1999) 943–945

[20] Mellen, C., Fröhlich, J., Rodi, W.: Karlruhe's mid-term report, LESFOIL: A Brite-Euram project. Technical report, Institut für Hydrodynamik, University of Karlsruhe, Germany (1999)

[21] Gough, T., Gao, S., Hancock, P., Voke, P.: Experiment and simulation of the tripped boundary layer on a flat plate: Comparative study. Technical Report ME-FD/95.40, Department of Mechanical Engineering, The University of Surrey, U.K. (1996)

[22] Nikiton, N., Nicoud, F., Wasistho, B., Squires, K., Spalart, P.: An approach to wall modeling in large-eddy simulations. Physics of Fluids A **12** (2000) 1629–1632

[23] Spalart, P., Jou, W.H., M.Strelets, Allmaras, S.: Comments on the feasability of LES for wings and on a hybrid RANS/LES approach. In Liu, C., Liu, Z., eds.: Advances in LES/DNS, First Int. conf. on DNS/LES, Louisiana Tech University, Greyden Press (1997)

[24] Shur, M., Spalart, P., Strelets, M., Travin, A.: Detached-eddy simulation of an airfoil at high angle of attack. In Rodi, W., Laurence, D., eds.: Engineering Turbulence Modelling and Experiments 4, Elsevier (1999) 669–678

[25] Travin, A., Shur, M., Strelets, M., Spalart, P.: Detached-eddy simulations past a circular cylinder. Flow Turbulence and Combustion **63** (2000) 293–313

[26] Peng, S.H., Davidson, L., Holmberg, S.: A modified Low-Reynolds-Number $k-\omega$ model for recirculating flows. ASME: Journal of Fluids Engineering **119** (1997) 867–875

[27] Yoshizawa, A.: Bridging between eddy-viscosity-type and second-order models using a two-scale DIA. In: 9th Int. Symp. on Turbulent Shear Flow. Volume 3., Kyoto (1993) 23.1.1—23.1.6

[28] Fureby, C.: Large eddy simulation of rearward-facing step flow. AIAA Journal **37** (1999) 1401–1410

[29] Piomelli, U.: High Reynolds number calculations using the dynamic subgrid-scale stress model. Physics of Fluids A **5** (1993) 1484–1490

[30] Mellen, C., Fröhlich, J., Rodi, W.: Large eddy simulation of the flow over periodic hills. In: 16th IMACS World Congress 2000, Lausanne, August 21-25. (2000)

[31] Smagorinsky, J.: General circulation experiments with the primitive equations. Monthly Weather Review **91** (1963) 99–165

[32] Temmerman, L., Leschziner, M., Mellen, C., Fröhlich, J.: Investigation of wall–function approximations and subgrid–scale models in large eddy simulation of separated flow in a channel with streamwise periodic constrictions (submitted). Int. J. Heat and Fluid Flow (2002)

[33] Mathey, F., Fröhlich, J., Rodi, W.: In preparation (2000)

[34] Temmerman, L., Leschziner, M.: Large eddy simulation of separated flow in a streamwise periodic channel constriction. In: The Second International Symp. on Turbulence and Shear Flow Phenomena. Volume 3., Stockholm (2001) 399–404

[35] Cabot, W., Moin, P.: Approximate wall boundary conditions in the large eddy simulation of high Reynolds number flow. Flow, Turbulence and Combustion **63** (1999) 269–291
[36] Nilsson, H., Davidson, L.: CALC-PVM: A parallel multiblock SIMPLE multiblock solver for turbulent flow in complex domains. Rept. 98/12, Dept. of Thermo and Fluid Dynamics, Chalmers University of Technology, Gothenburg (1998)
[37] FLUENT Newsletter: Collaboration with the Silicon Graphics Delivers Breakthrough Parallel Scaling for High-End CFD Computations. Volume 6:2. (1997)
[38] Davidson, L., Farhanieh, B.: CALC-BFC: A finite-volume code employing collocated variable arrangement and cartesian velocity components for computation of fluid flow and heat transfer in complex three-dimensional geometries. Rept. 92/4, Dept. of Thermo and Fluid Dynamics, Chalmers University of Technology, Gothenburg (1992)
[39] Nilsson, H.: A parallel multiblock extension to the CALC-BFC code using PVM. Rept. 97/11, Dept. of Thermo and Fluid Dynamics, Chalmers University of Technology, Gothenburg (1997)
[40] Hellström, T., Davidson, L.: A multiblock-moving mesh extension to the calc-bfc code. Rept. 93/3, Dept. of Thermo and Fluid Dynamics, Chalmers University of Technology, Gothenburg (1993)
[41] Ferziger, J., Peric, M.: Computational Methods for Fluid Dynamics. Springer-Verlag, Berlin (1996)
[42] Mathur, S., Murthy, J.: Pressure-based method for unstructured methods. Numerical Heat Transfer. Part B **31** (1997) 195–216
[43] Prasad, R., S.E., K.: Cold flow simulations of a bluff-body flame stabiliser. Fluent Inc. (1998)
[44] 6th ERCOFTAC/IAHR/COST Workshop on Refined Flow Modelling. In: June 6-7, Delft University of Technology, Delft, The Netherlands. (1997)
[45] Martinuzzi, R., Tropea, C.: The flow around a surface-mounted prismatic obstacle placed in a fully developed channel flow. J. Fluid Engng. **115** (1993) 85–92
[46] Breuer, M.: LES of the flow past bluff body: Numerical and modelling aspects. In: DNS and LES of Complex Flows: Numerical and Modelling Aspects, Twente University, The Netherlands (1997) 14–21
[47] Rhie, C., Chow, W.: Numerical study of the turbulent flow past an airfoil with trailing edge separation. AIAA Journal **21** (1983) 1525–1532
[48] Issa, R.: Solution of Implicitly Discretised Fluid Flow Equations by Operator-Splitting. J. Comp. Physics **62** (1986) 40–65
[49] Yoshizawa, A.: Statistical theory for compressible shear flows with the application of subgrid modelling. Physics of Fluids A **29** (1986) 2152–2163
[50] Dahlström, S., Davidson, L.: Large eddy simulations of the flow around an Aerospatiale A-airfoil. Proceedings of ECCOMAS 2000, Barcelona, 11–14 September (2000)
[51] Dahlström, S., Davidson, L.: Large eddy simulations of the flow around an airfoil. AIAA Paper 2001-0425, Reno (2001)
[52] Davidson, L., Peng, S.H.: A hybrid LES–RANS model based on a one-equation SGS model and a two-equation $k - \omega$ model. In Lindborg, E., Johansson, A., Eaton, J., Humphrey, J., Kasagi, N., Leschziner, M., Sommerfeld, M., eds.:

The Second International Symp. on Turbulence and Shear Flow Phenomena. Volume 2., Stockholm (2001) 175–180

[53] Arina, R., Ceresola, N., Piant, P.: Application of a two equations turbulence model to the numerical prediction of unsteady flow past an airfoil. In: Proc. 20th ICAS Congress. (1996)

[54] Johansson, A., Wallin, S.: A new explicit algebraic Reynolds stress model. In: Advances in Turbulence VI, Netherlands, Kluwer Academic Publishers (1996) 31–34

[55] Spalart, P., Allmaras, S.: A one equation turbulence model for aerodynamic flows. AIAA paper 92-0439, Reno, NV (1992)

[56] Weber, C., Ducros, F.: Large-eddy and Reynolds averaged Navier–Stokes simulations of turbulent flow over an airfoil. Int. J. Comput. Fluid Dyn. **13** (2000) 327–355

[57] Weber, C.: Développement de méthodes implicites pour les équations de navier-stokes moyennées et la simulation des grandes échelles: application à l'aérodynamique externe. thèse, INPT, Toulouse, France (1998)

[58] Weber, C., Ducros, F., Corjon, A.: Large-eddy simulation of complex turbulent flows. In: AIAA Paper 98-2651. (1998)

[59] Ducros, F., Comte, P., Lesieur, M.: Large-eddy simulation of transition to turbulence in a boundary layer developing spatially over a flat plate. Journal of Fluid Mechanics **326** (1995) 1–36

[60] Nicoud, F., Ducros, F.: Subgrid-scale stress modelling based on the square of the velocity gradient tensor. Flow, Turbulence and Combustion. **62** (1999) 183–200

[61] Caruelle, B.: Simulation d'ecoulements instationnaires turbulents en aérodynamique: application à la prédistion du phénomène de tremblement. thèse de l'INPT (2000)

[62] Caruelle, B., Ducros, F.: Towards unsteady simulations of separation of boundary layer. in Proceedings of 1th TSFP congress (Santa Barbara, 1999) pp. 1155–1160

[63] Ducros, F., Soulères, C., Laporte, F., Moinat, P., Weber, C., Guinot, V.: High-order skew-symmetric Jameson schemes for unsteady compressible dlow. In: ERCOFTAC joint Workshop on Direct and Large-Eddy Simulation, Cambridge, May 12-14. (1999)

[64] Chalot, F., Mallet, M., Ravachol, M.: A comprehensive finite element Navier-Stokes solver for low- and high-speed aircraft design. AIAA Paper 94-0814, Reno, Nevada (1994)

[65] Hughes, T., Franca, L., Mallet, M.: A new finite element formulation for computational fluid dynamics: VI convergence analysis of the generalized SUPG formulation for linear time dependent multidimensional advective diffusive systems. Computer Methods in Applied Mechanics and Engineering **63** (1987) 97–112

[66] Johan, Z., Hughes, T., Shakib, F.: A globally convergent matrix-free algorithm for implicit time-marching schemes arising in finite element analysis in fluids. Computer Methods in Applied Mechanics and Engineering **87** (1991) 281–304

[67] Johnson, C.: Finite elements methods for flow problems. In: Proceedings in Agard report 787 for "Special course on unstructured grid method for advection dominated flows". (1992)

[68] Shakib, F.: Finite element analysis of the compressible Euler and Navier-Stokes equations. PhD thesis, Division of Applied Mechanics, Stanford University (1989)
[69] Hughes, T., Mallet, M., Mizukami, A.: A new finite element formulation for computational fluid dynamics: II beyond SUPG. Computer Methods in Applied Mechanics and Engineering **54** (1986) 341–355
[70] Johan, Z.: Data parallel finite element techniques for large-scale computational fluid dynamics. PhD thesis, Stanford University (1992)
[71] Spalart, P., Allmaras, S.: A one–equation turbulence model for aerodynamic flows. La Recherche Aérospatiale **1** (1994) 5–21
[72] Chen, H., Patel, V.: Near-wall turbulence models for complex flows including separation. AIAA Journal **26** (1988) 641–648
[73] Kasbarian, C.: Étude numérique pour les problèmes d'advection-diffusion – Application à la modélisation d'écoulements turbulents incompressibles. PhD thesis, Université de Lyon I (1995)
[74] Marquez, B.: Simulation des Grandes Échelles d'Écoulements Compressibles par des Méthodes Éléments Finis. PhD thesis, Institut National Polytechnique de Toulouse (1999)
[75] Weber, C.: Développement de méthodes implicites pour les équations de Navier-Stokes moyennées et à la simulation des grandes échelles : Application à l'aérodynamique externe. PhD thesis, Institut National Polytechnique de Toulouse (1998)
[76] Cokljat, D.: LES of flow around A-Airfoil Part I - RANS results and initial les simulation, internal report, LESFOIL Project. Technical report, Fluent, Sheffield, UK (1999)
[77] Zhu, J.: A low diffusive and oscillation-free convection scheme. Communications in Applied Numerical Methods **7** (1991) 225–232
[78] Davidson, L.: LESFOIL: A european project on large eddy simulations around a high-lift airfoil at high Reynolds number. In: ECCOMAS 2000, European Congress on Computational Methods in Applied Sciences and Engineering, 11-14 September, Barcelona, Spain (2000)
[79] Péchier, M.: Prévision numériques de l'effet Magnus pour des configurations de munitions. PhD thesis, Thèse de l'université de Poitiers (1999)
[80] Péchier, M., Guillen, P., Gayzac, R.: Magnus effect over finned projectiles. J. of Aircraft and Rockets **38** (2001) 542–549
[81] Lenormand, E., Sagaut, P., Ta Phuoc, L., Comte, P.: Subgrid-scale models for large-eddy simulation of compressible wall bounded flows. AIAA J. **38** (2000) 1340–1350
[82] Lenormand, E., Sagaut, P., Ta Phuoc, L.: Large-eddy simulation of subsonic and supersonic channel flow at moderate reynolds number. Int. J. Numer. Meth. Fluids **32** (2000) 369–406
[83] David, E.: Modélisation numérique d'écoulements compressibles:une approche instationnaire. Thèse de docteur-ingénieur, INPG (1993)
[84] Sagaut, P.: Large-eddy simulation of incompressible flows. An introduction. Springer-Verlag (2001)
[85] J.P. Boris, F.F. Grinstein, E.O.E., Kolbe, R.: New insights into large eddy simulation. Fluid Dynam. Res. **10** (1992) 199
[86] Edwards, J.R., Liou, M.S.: Low-diffusion flux-splitting methods for flows at all speeds. AIAA J. **36** (1998) 1610–1617

[87] Mary, I., Sagaut, P., Deville, M.: An algorithm for unsteady viscous flows at all speeds. Int. J. Numer. Meth. Fluids **34** (2000) 371–401

[88] Mary, I.: Méthode de Newton approchée pour le calcul d'écoulements instationnaires comportant des zones à très faibles nombres de Mach. PhD thesis, Thèse de l'université Parix XI Orsay (1999)

[89] Ducros, F.: private communication (2000)

[90] Gleyzes, C.: Opération décrochage - résultats de essais à la soufflerie F2. Technical Report RT-DERAT 55/4004, ONERA (1988)

[91] Na, Y., Moin, P.: Direct numerical simulation of a separated turbulent boundary layer. Journal of Fluid Mechanics **374** (1998) 379–405

[92] Alam, M., Sandham, N.: Direct numerical simulation of short laminar separation bubbles with turbulent reattachment. J. Fluid Mech. **410** (2000) 1–28

[93] Sohankar, A., Norberg, C., Davidson, L.: Simulation of unsteady three-dimensional flow around a square cylinder at moderate Reynolds numbers. Physics of Fluids A **11** (1999) 288–306

[94] Temmerman, L., Leschziner, M., Ashworth, M., Emerson, D.: Les applications on parallel systems. In: Proceedings of Parallel CFD 2000. (2000)

[95] Pauley, L., Moin, P., Reynolds, W.: The structure of two-dimensional separation. J. Fluid Mech. **220** (1990) 397–411

[96] Dahlström, S., Davidson, L.: Chalmers 30-month report, LESFOIL: a Brite-Euram project. Technical report, Chalmers, Goteborg, Sweden (2000)

[97] Mellen, C., Fröhlich, J., Rodi, W.: Ifh 30-month report, LESFOIL: a Brite-Euram project. Technical report, University of Karlsruhe, Karlsruhe, Germany (2000)

[98] Werner, H., Wengle, H.: Large-eddy simulation of turbulent flow over and around a cube in a plane channel. In: Turbulent Shear Flows 8, Springer-Verlag (1991) 155–168

VII. ADDRESSES OF PARTNERS

VII. ADDRESSES OF PARTNERS

1. Chalmers University
Prof. L. Davidson (coordinator)
Dept. of Thermo and Fluid Dynamics
Chalmers University of Technology
S-412 96 Gothenburg
SWEDEN
E-mail: lada@tfd.chalmers.se
http://www.tfd.chalmers.se/~lada

2. Alenia
Dr. N. Ceresola
Alenia Aerospazio – Divisione Aeronautica
C.So Marche, 41
I-101 46 Torino
ITALY
E-mail: nceresola@aeronautica.alenia.it
http://www.alenia.it

3. CERFACS
Dr. F. Ducros[a]
CERFACS
Av. G. Corilolis, 42
F-310 57 Toulouse Cedex
FRANCE
http://www.cerfacs.fr

[a] Present address: DTP/SMTH/LDTA, CEA-Grenoble, France

4. Dassault
Dr. M. Mallet
Dassault Aviation
DGT/DTA/MOD
78, Quai Marcel Dassault
F-922 14 St Cloud
FRANCE
E-mail: michel.mallet@dassault-aviation.fr
http://www.dassault.fr

5. Fluent Europe Ltd
Dr. D. Cokljat
Fluent Europe Ltd
Sheffield Airport Business Park
Europa Link
Sheffield S9 1XU
UK
E-mail: davor@fluent.co.uk
http://www.fluent.com/worldwide/europe/about/index.htm

6. Karlsruhe University
Prof. W. Rodi
Institut für Hydrodynamik
Universität Karlsruhe
Kaiserstrasse 12
P.O. Box 6980
D-76128 Karlruhe
GERMANY
E-mail: rodi@ifh.bau-verm.uni-karlsruhe.de
http://www-ifh.bau-verm.uni-karlsruhe.de/ifh/

7. ONERA
Dr. P. Sagaut
Av. de la Division Leclerc, no. 29
ONERA
BP 72
F-923 22 Châtillon Cedex
FRANCE
E-mail: sagaut@onera.fr
http://www.onera.fr

8. The University of Surrey
Dr. P. Voke
Dept. of Mechanical Engineering
The University of Surrey
Guildford GU2 5XH
GREAT BRITAIN
E-mail: p.voke@surrey.ac.uk
http://www.surrey.ac.uk/eng/staff.htm

9. UMIST
Prof. Michael A Leschziner[a]
Dept. of Mechanical Engineering
UMIST
P.O. Box 88
Manchester M60 1QD
GREAT BRITAIN

[a] Present address: Imperial College of Science, Technology and Medicine, Aeronautics Department, Prince Consort Rd., London SW7 2BY, GREAT BRITAIN, E-mail: mike.leschziner@ic.ac.uk,http://www.ae.ic.ac.uk/staff/leschziner/index.html

Druck: Strauss Offsetdruck, Mörlenbach
Verarbeitung: Schäffer, Grünstadt

Notes on Numerical Fluid Mechanics and Multidisciplinary Design

Available Volumes

Volume 82: E. H. Hirschel (ed.): Numerical Flow Simulation III - CNRS-DFG Collaborative Research Programme, Results 2000-2002. ISBN 3-540-44130-1

Volume 81: W. Haase, V. Selmin, B. Winzell (eds.): Progress in Computational Flow-Structure Interaction - Results of the Project UNSI, supported by the European Union 1998-2000. ISBN 3-540-43902-1

Volume 80: E. Stanewsky, J. Délery, J. Fulker, P. de Matteis (eds.): Drag Reduction by Shock and Boundary Layer Control - Results of the Project EUROSHOCK II, supported by the European Union 1996-1999. ISBN 3-540-43317-1

Volume 79: B. Schulte-Werning, R. Grégoire, A. Malfatti, G. Matschke (eds.): TRANSAERO - A European Initiative on Transient Aerodynamics for Railway System Optimisation. ISBN 3-540-43316-3

Volume 78: M. Hafez, K. Morinishi, J. Periaux (eds.): Computational Fluid Dynamics for the 21st Century. Proceedings of a Symposium Honoring Prof. Satofuka on the Occasion of his 60th Birthday, Kyoto, Japan, 15-17 July 2000. ISBN 3-540-42053-3

Volume 77: S. Wagner, U. Rist, H.-J. Heinemann, R. Hilbig (eds.): New Results in Numerical and Experimental Fluid Mechanics III. Contributions to the 12th STAB/DGLR Symposium Stuttgart, Germany 2000. ISBN 3-540-42696-5

Volume 76: P. Thiede (ed.): Aerodynamic Drag Reduction Technologies. Proceedings of the CEAS/DragNet European Drag Reduction Conference, 19-21 June 2000, Potsdam, Germany. ISBN 3-540-41911-x

Volume 75: E.H. Hirschel (ed.): Numerical Flow Simulation II. CNRS-DFG Collaborative Research Programme, Results 1998-2000. ISBN 3-540-41608-0

Volume 74: D. Henry, A. Bergeon (eds.): Continuation Methods in Fluid Dynamics. Contributions to the ERCOFTAC/EUROMECH Colloquium 383, Aussois, France, 6.-9. September 1998. ISBN 3-540-41556-4

Volume 73: M. Griebel, S. Margenov, P. Yalamov (eds.): Large-Scale Scientific Computations of Engineering and Environmental Problems II. Proceedings of the Second Workshop on „Large-Scale Scientific Computations", Sozopol, Bulgaria, June 2-6, 1999. ISBN 3-540-41551-3

Volume 72: W. Nitsche, H.-J. Heinemann, R. Hilbig (eds.): New Results in Numerical and Experimental Fluid Mechanics II. Contributions to the 11th AG STAB/DGLR Symposium Berlin, Germany 1998. ISBN 3-540-41552-1

Volume 70: W. Hackbusch, G. Wittum (eds.): Numerical Treatment of Multi-Scale Problems. Proceedings of the 13th GAMM-Seminar, Kiel, January 24-26, 1997. ISBN 3-540-41562-9

Volume 69: D. P. Hills, M. R. Morris, M. J. Marchant, P. Guillen (eds.): Computational Mesh Adaptation. ECARP - European Computational Aerodynamics Research Project. ISBN 3-540-41548-3

Volume 67: C. H. Sieverding, G. Cicatelli, J. M. Desse, M. Meinke, P. Zunino: Experimental and Numerical Investigation of Time Varying Wakes behind Turbine Blades. Results of the Industrial and Material Technology Aeronautics Research Project AER2-CT92-0048, 1992-1996. ISBN 3-540-41546-7

Volume 66: E. H. Hirschel (ed.): Numerical Flow Simulation I. CNRS-DFG Collaborative Research Programme. Results 1996-1998. ISBN 3-540-41540-8

Volume 64: R. Friedrich, P. Bontoux (eds.): Computation and Visualization of Three-Dimensional Vortical and Turbulent Flows. Proceedings of the Fifth CNRS-DFG Workshop on Numerical Flow Simulation, München, Germany, December 6 and 7, 1996. ISBN 3-540-41557-2

Volume 63: M. Fiebig, N. K. Mitra (eds.): Vortices and Heat Transfer. Results of a DFG-Supported Research Group. ISBN 3-540-41558-0

Volume 62: M. Griebel, O. Iliev, S. D. Margenov, P. S. Vassilevski (eds.): Large-Scale Scientific Computations of Engineering and Environmental Problems. Proceedings of the First Workshop on „Large-Scale Scientific Computations“ Varna, Bulgaria, June 7-11, 1997. ISBN 3-540-41560-2

Volume 57: H. Deconinck, B. Koren (eds.): Euler and Navier-Stokes Solvers Using Multi-Dimensional Upwind Schemes and Multigrid Acceleration. Results of the BRITE/EURAM Projects AERO-CT89-0003 and AER2-CT92-00040, 1989-1995. ISBN 3-540-41535-1

GPSR Compliance

The European Union's (EU) General Product Safety Regulation (GPSR) is a set of rules that requires consumer products to be safe and our obligations to ensure this.

If you have any concerns about our products, you can contact us on ProductSafety@springernature.com

In case Publisher is established outside the EU, the EU authorized representative is:

Springer Nature Customer Service Center GmbH
Europaplatz 3
69115 Heidelberg, Germany

Batch number: 10755292

Printed by Printforce, the Netherlands